Chiara Mannelli

The Ethics of Rapid Tissue Donation (RTD)

Constructing a Formal and Substantial
Informed Consent Process

 Springer

Chiara Mannelli
Department of Philosophy and Education Sciences
University of Turin
Turin, Italy

Candiolo Cancer Institute, FPO-IRCCS
Candiolo, TO, Italy

ISSN 2662-9186   ISSN 2662-9194 (electronic)
The International Library of Bioethics
ISBN 978-3-030-67200-3   ISBN 978-3-030-67201-0 (eBook)
https://doi.org/10.1007/978-3-030-67201-0

This Springer imprint is published by the registered company Springer Nature Switzerland AG
The registered company address is: Gewerbestrasse 11, 6330 Cham, Switzerland

Acknowledgements

I am especially grateful to Maurizio Mori, whose insightful guidance has been pivotal from the very beginning.

During the long journey that stands behind this book, I have been lucky enough to have met outstanding professionals that deeply inspired my work. Among them, I am particularly indebted to Anna Sapino for her unconditional support and scientific oversight; to Giancarlo Di Vella, Oreste Buonomo and Maria Novella Luciani for their special engagement; to Daniela Gramaglia for being part of this project; and to those who shared with me opinions and other background material along the way.

A special thank goes to Stephen Pigney for his invaluable contribution, to the anonymous reviewers for their sharp suggestions, and to the University of Turin, FINO PhD program for funding this project.

Lastly, I owe all my gratitude to the Patient without whose generosity this work would never have started.

Inscription:
To *M.P., F.E.G. & T.*

Introduction

Cancer is the second leading cause of death in developed countries, after heart diseases. Although cancer can be partially prevented by avoiding risk factors and implementing existing evidence-based prevention strategies, oncology research aimed at developing effective treatments plays a key role in preventing cancer deaths. Due to advances in knowledge and treatments, the chances of surviving cancer are increasingly improving if the disease is diagnosed early and treated correctly. Research has shed light on different aspects of cancer that have made it possible to partially understand the biological mechanisms that may cause cancer onset and to determine its consequent development and resistance to some treatments. Such understanding is the cornerstone of the implementation of preventive measures, and it is driving the development of promising cancer therapies.

Prevailing theory, which was formulated in the middle of the twentieth century, interprets cancer as a set of about 200 diseases characterized by an abnormal cell growth that is released from the normal control mechanisms of the organism. While there is normally a balance between proliferation and programmed cell death, mutations in DNA leading to cancer are responsible for the destruction of these ordered processes. The study of the molecular structure of tumors, in constant variation, is the research field with the most promising outlook for future clinical developments.

Research in genomics is making possible a targeted and personalized approach to the diagnosis and treatment of different diseases. This approach, which is referred to as "precision medicine," is based on a model of healthcare delivery that relies on individual variability in genes, environment, and lifestyle. By understanding this variability, it is possible to classify individuals into subpopulations that differ in their susceptibility to a particular disease or in their response to a specific treatment. This model enables healthcare decisions to be guided toward the most effective care and treatment for a given patient based on their personal profile.

The development of precision medicine has a wide array of important implications. Clinical implications are at the forefront, but they are not the only ones. Precision medicine raises various ethical issues. The ability to sequence the entire genome of an individual patient presents researchers with a golden opportunity to advance research in the field by exploring the molecular landscape of genes associated with human diseases. At the same time, however, this opportunity depends on an unprecedented amount of private, and identifiable (or reidentifiable) data being made available. The management of and access to this data require adequate policies that take into account research needs on the one hand, while safeguarding patients on the other. Analyzing the ethical issues raised by the application of omics technologies within the clinical field involves, among other things, a consideration of how the sharing of genetic information and genetic data may impact individuals, their families, and their communities. The use of genetic information in research is key to advances in precision medicine and holds unique opportunities for the development of new treatments, but the pursuit of this opportunity can in no way be detrimental to patients' rights. Thus, the development of an ethical framework to support policies relating to informed choices and to the responsible use of personal information assumes unprecedented importance to the goal of fostering the development of science while protecting the rights of patients.

Within oncology, precision medicine uses a knowledge of the molecular landscape of tumors in order to tailor medical treatment to the individual characteristics of each patient. The development and applicability of this kind of targeted approach in the field of oncology heavily relies on the availability of tumor tissues to be used for research aims. The procurement of adequate cancer tissues for research using traditional biopsies in living patients is problematic because of the limited quality and quantity of the retrieved tissues; however, postmortem retrieval of tissues represents a far more promising opportunity for collection of suitable tissues.

Rapid Tissue Donation (RTD) in oncology is an advanced procedure that involves the procurement of "fresh" cancer tissue ideally within 2–6 hours of the death of a patient (Lindell et al. 2006). In this technique, the restricted time frame in which the material is retrieved from the patient is essential for preserving the molecular properties of collected tissues. RTD enables researchers to extract a significant quantity of high-quality tumor tissue from both the primary site of the tumor and from all its derived metastases; in addition, unaffected tissues can be retrieved in order to study the molecular composition of the tissue. Fresh tissues obtained via RTD offer research opportunities that are not available using other techniques: it is not feasible to make a similar collection of tissues from living patients, and tissues previously collected and stored using preservation systems fall significantly below the quality of tissue that can be obtained from RTD.

RTD represents a significant chance to investigate cancer onset and development in order to translate acquired knowledge into new treatment options for future cancer patients. However, despite the high promise of this procedure, its implementation in the clinical setting raises challenging issues associated with the informed consent process. First, RTD tissue retrieval takes place *after* death, when the need for an informed consent has been questioned. Moreover, because the quality and quantity of retrieved tissues offer insights that enable research on new treatments, the need for an informed consent process to govern RTD has been questioned on the grounds that the potential benefits to society from this research outweigh any requirement for informed consent. The lack of consensus on the need for an informed consent procedure to regulate this form of tissue collection hampers its implementation within the clinical setting, and thus also creates barriers to the development of cancer research.

The aim of this book is to analyze the ethical quandaries raised by the informed consent process for RTD, in particular by building on a specific interpretation of the notion of informed consent that envisions the coexistence of two senses associated with this concept. Drawing on the work of R. R. Faden and T. L. Beauchamp, I consider informed consent to comprise both a *substantial* meaning and a *formal* meaning. In its *substantial* sense, informed consent is a specific form of autonomous authorization conferred by a patient or a subject. This authorization lies at the heart of this sense of informed consent: in authorizing, patients both assume responsibility for what they have authorized and transfer it to another's authority in order for what they have authorized to be implemented. At the same time, there is an institutional set of rules and policy governing this act, and it is to these that the *formal* sense of informed consent refers. In its *formal* sense, informed consent entails a legally effective authorization from a patient, and the effectiveness of this act is determined by its compliance with the rules and requirements that define a specific institutional practice in healthcare and research (Faden and Beauchamp 1986, p. 280). In this second sense, informed consent is policy-oriented and reflects the conformity to the social rules of consent that require medical professionals to seek and obtain valid consent from patients before carrying out medical or research interventions. This public practice is regulated by specific norms and institutional settings that are intended to define when and if a certain consent or refusal provided by a patient is valid.

In this framework, the relationship between the two senses of informed consent is deep: they are separate and do not overlap, yet at the same time they are closely tied together. Policies governing the *formal* sense of the term should be formulated to conform to the standards of the *substantial* sense; in other words, institutional requirements for *formal* informed consent should be intended to maximize the likelihood that the conditions of informed consent in its *substantial* sense will be satisfied. The autonomy-based model of informed consent in its *substantial* sense should function as a standard for informed consent in a *formal* sense (Faden and Beauchamp 1986, p. 284).

However, a significant discrepancy emerges from the relationship between these two senses of informed consent. It may happen that an informed consent given in the first sense does not represent a legally effective informed consent in the second sense. This means that an informed consent obtained under institutional criteria (envisioned by the *formal* sense) may fail to conform to the autonomy-based model (the *substantial* sense), and vice versa. Thus, informed consent obtained according to its *substantial* sense may, in certain cases, not be effective in its *formal* sense, and, *mutatis mutandis*, informed consent compliant with its *formal* sense may not be effective in its *substantial* sense.

This dual-sense structure of informed consent provides the lens through which the unprecedented issues raised by informed consent for RTD will be analyzed. The structure is used to facilitate the discussion and analysis of the two key areas in this book, namely, whether there is a need for informed consent to RTD, and, if so, how to implement an RTD informed consent procedure.

The framework offered by the *formal* sense of informed consent, intended as a policy-oriented authorization defined by institutional settings, will be considered in relation to the first key area of analysis, which concerns whether there is a need for informed consent to govern RTD. This need has been challenged by various commentators on different grounds; in this book, arguments both for and against informed consent will be considered, but it will ultimately be argued that there *is* a need for an informed consent to govern RTD.

After making the case for the need for an informed consent process to regulate RTD, my analysis will turn to the ethical quandaries raised by the practical implementation of informed consent. In relation to RTD, how should informed consent be structured and collected in order to encourage an autonomous authorization intended in the *substantial* sense of the term? And how should this autonomy be formalized in the *formal* sense of informed consent? The innovative nature of RTD makes the informed consent process challenging: as well as raising some ethical issues common to standard oncological research and donation procedures, RTD also poses novel issues that have no precedents in the medical setting. In addressing and analyzing these questions and issues surrounding informed consent within the RTD setting, it will be proposed that there should be a dedicated informed consent procedure in which a specifically trained RTD ethicist should play a central role.

By viewing medical issues relating to informed consent in oncology through an ethical lens, the analysis blends traits of abstract philosophy with concrete cancer-related aspects. As a result, the book is situated at the intersection of various interests: it is suitable for readers fascinated by ethical reasoning as well as for those with a medical background. To achieving this aim of meeting the needs of a heterogeneous group of readers, I have attempted to present the analysis in such a way that it will be clear both to those who are new to many of the areas of discussion and to those who are highly familiar with the topic.

The structure of the book, composed of eight chapters in addition to the introduction and conclusion, reflects the inter- and multidisciplinary nature of the subject at stake.

Chapter 1 introduces RTD, and the wider context of precision medicine in oncology, in more detail. It outlines, from a technical point of view, the potential research benefits of collecting cancer tissues immediately after the death of the patient.

In Chapter 2, I turn my attention to the issue of informed consent, which lies at the heart of the relationship between patients and physicians. I consider how this relationship has evolved and the implications of this evolution for the meaning and boundaries of informed consent. I then analyze the two senses of informed consent—the *substantial* sense and the *formal* sense—which constitute the framework within which I will later discuss informed consent for RTD.

Chapter 3 opens with a presentation of Beauchamp and Childress's analysis of the four principles that constitute the foundation of biomedical reflection, namely, the principles of autonomy, beneficence, non-maleficence, and justice. This framework serves as a starting point to explore in more depth the issues associated with informed consent for RTD. First, arguments against the need for informed consent to govern postmortem procedures will be introduced. I then survey some important positions advocating a regime of conscription. Finally, I address arguments for the need for an informed consent to govern postmortem procedures.

Having surveyed various philosophical positions for and against informed consent in the third chapter, in Chapter 4 I develop an argument for the need for an informed consent procedure to govern RTD. I begin by introducing the concept of the "once-alive" as a way of thinking about the dead as individuals who had values, wishes, and preferences in life that should be honored after death. In this way, I argue that there is a compelling basis for requiring informed consent based on two principles: (1) it is crucial to honor, even after death, the wishes (if any) that individuals have expressed during their lives, (2) provided that these preferences do not jeopardize other living individuals. After setting out this argument, I then discuss why positions that advocate no need for an informed consent to regulate postmortem procedures should be rejected.

Chapter 5 delves into the relevance of informed consent in the medical field, with a specific focus on its role in the research setting. I discuss the main regulations and codes that govern informed consent within medical research, and I analyze information, comprehension, and voluntariness as the basic features of informed consent in the research setting. These requirements represent the basis to consider the issues raised by informed consent in research through a comparison between oncology research and RTD.

Chapter 6 bridges the ethical and medical dimensions by introducing the figure of the clinical ethicist. The aim of this chapter is to underline the value of an ethical perspective for viewing the issues that emerge on a daily basis in the medical setting. The value of this perspective is particularly relevant within the context of RTD whose innovative traits raise unprecedented and challenging ethical issues. I also argue for the need for an RTD ethicist—namely, an ethicist with specific training in RTD programs—to adequately support patients, family, and medical staff throughout the RTD process, particularly in relation to informed consent.

Chapter 7 outlines an informed consent process, intended in both the *formal* and *substantial* senses of the term, for RTD. This chapter addresses the practical aspects of RTD informed consent and their relevant ethical implications by considering questions such as who should be in charge of presenting RTD as an option to the patient. I propose a three-phase structure for informed consent that is designed to promote and encourage, as far as possible, a choice that reflects a patient's preferences. Within this chapter, aspects pertaining to informed consent for vulnerable populations are also discussed.

The structure for an RTD informed consent process presented in the seventh chapter paves the way for the ethical analysis in Chapter 8. This analysis retains a practical approach. Challenging issues associated with informed consent for RTD are addressed, such as consent withdrawal, the role of families in the informed consent procedure, how collected samples may be identified (or reidentified), and the various possible uses for tissues in research. Within this scenario, the role of the RTD ethicist is vital for a thorough informed consent valid in both the *formal* and *substantial* senses of the term.

The ultimate objective of this book is to set out an informed consent for RTD intended in its *formal* sense—and hence a legally recognized path for a cancer patient to rely on when deciding whether to donate their tissues after death—but one whose requirements are modelled according to the *substantial* sense of informed consent—namely, one that honors the autonomous authorization of the patient.

References

Faden, R., and Beauchamp, T. L. 1986. *A History and Theory of Informed Consent*. Oxford: Oxford University Press.

Lindell, K. O. et al. 2006. Lessons from our patients: development of a warm autopsy program. *PLoS Medicine* 3 (7): e234.

Contents

Chapter 1
Rapid Tissue Donation (RTD) for Oncology Research

Abstract Research in genomics and "omics" technologies is making possible "precision medicine", a targeted approach to the diagnosis and treatment of diseases. The goal of precision medicine, and one that holds great promise in oncology, is to treat patients with drugs that target the specific genetic mutations in their tumors. The research use of cancer tissues aims to unravel the mechanisms of cancer onset and evolution. To this end, collection of cancer tissues after the death of a cancer patient represents a unique opportunity for researchers. Rapid Tissue Donation (RTD) is an advancing oncology procedure that involves the procurement of "fresh" tissue within 2–6 h following the death of a cancer patient. This window of time is ideal to preserve the high quality of tissues retrieved. Whereas traditional tissue biopsies provide researchers with a limited amount of material that is often suboptimal for research needs, RTD offers the chance to overcome these barriers.

1.1 Precision Medicine

The next generation of medicine is upon us. Advances in genomics and other "omics" technologies over the past decade have yielded new tools to evaluate disease susceptibility and prognosis, and they offer unprecedented opportunities to individualize therapy. New medicines are increasingly targeted to specific patient populations and have enriched the therapeutic approach to treatment of a wide range of pathologies, among which are cancer, chronic infections, and rare diseases.

This targeted approach to the diagnosis and treatment of different diseases is known as "precision medicine", an emerging model for healthcare delivery that relies on individual variability in genes, as well as on environment and lifestyle.[1] According to the definition given in 2008 by the President's Council of Advisors on Science and Technology (National Research Council 2011), precision medicine is the

---

[1] Among the literature on precision medicine, see: Aronson and Rehm (2015), Delaney et al. (2016), Dzau and Ginsburg (2016), Jameson and Longo (2015), Khoury and Galea (2016), Khoury et al. (2016), McCarthy et al. (2013), Mirnezami et al. (2012), National Research Council (2011), Phillips et al. (2017), Vargas and Harris (2016).

---

tailoring of medical treatment to the individual characteristics of each patient to classify individuals into subpopulations that differ in their susceptibility to a particular disease or their response to a specific treatment.

The advent of precision medicine has disrupted the entire medical field. Its introduction of new technologies has revolutionized the progress and potential of medicine, impacting on research, prevention, diagnosis, and care. In the pre-genomics era, most drugs were traditionally tested, approved, and developed according to their effects on a general population of patients affected by the same disease and a certain diagnosis. In the conventional approach of the "same drug fits all with the same disease", clinical trials for the development of new drugs or a new combination of drugs showed the average response of the patients' cohort to the treatment under investigation. However, this average response can obscure the fact that some patients might show an extremely positive response whereas others show no response at all. The DNA sequence between any two individuals (apart from identical twins) is approximately 99.9% identical, but the 0.1% difference is "medically significant": "enclosed within this small percentage of difference lie the clues to hereditary susceptibility to virtually all diseases" (Jain 2009, p. 1). It is in this 0.1% difference that can be found the reasons why responses to treatment differ among patients affected by the same disease.

Precision medicine holds great promise in oncology because it involves tailoring treatments that are best suited to each individual patient. A key part of such treatments is the integration of new technologies within the clinical care of patients. However, this does not mean creating drugs or medical devices that are unique to a patient; rather, it involves the ability to group patients according to their susceptibility to a particular disease, to the biology or prognosis of those diseases they may develop, or to their response to a specific treatment.[2] The potential created by this shift in treatment has dramatic implications along both medical and ethical dimensions. Preventive or therapeutic interventions can be concentrated on those who will benefit; and harm and side effects can be reduced for those who will not benefit. This model allows healthcare decisions to be guided toward the most effective treatment for a given patient, thereby improving care quality while reducing the need for unnecessary diagnostic testing and therapies.

Consequently, precision medicine is based on the study of a specific profile in order to choose the optimal targeted therapy. This approach will enable doctors and researchers to predict more accurately which treatment and prevention strategies for a particular disease will work in which groups of people, based on a genetic understanding of their disease.[3] The underlying concept of precision medicine, in which

---

[2] Among the oncological literature on this, see: Buettner et al. (2013), Garraway (2013), Garraway and Lander (2013), Lolkema et al. (2013), MacConaill (2013), Meric-Bernstam et al. (2013), Sleijfer et al. (2013), Van Allen et al. (2013).

[3] Among the literature on genome sequencing and genetic understanding, see: Cooper (2015), Denny et al. (2013), Gahl et al. (2016), Green et al. (2013), Hunter et al. (2016), Landrum et al. (2016), Ledford (2015), Minikel et al. (2016), Richards et al. (2015), Ritter et al. (2016), Schughart et al. (2013), Starita et al. (2015), Willig et al. (2015), Yamamoto et al. (2014), Yang et al. (2014).

healthcare is individually tailored on the basis of a person's genes, lifestyle, and environment, will facilitate the development and production of safer and more effective medicines with reduced occurrence of side effects. In the long term, individuals will have the opportunity to manage their health with bespoke approaches based on their genetic profile.

With the advent of genome sequencing, precision medicine has become a promising perspective from which to observe diseases. A gene can be defined for practical purposes as a "physical and functional unit of heredity, which carries information from one generation to the next", while, in molecular terms, it is the "entire DNA sequence including exons, introns, and noncoding transcription control regions that are necessary for production of a functional protein or RNA" (Jain 2009, p. 18). The sequencing of the human genome has provided researchers with crucial information for studying the genetic dimension of diseases. Within this framework, the identification of human genes associated with their regulatory locations enables researchers to study diseases in intimate detail in order to better understand their functioning, onset, and development. Advances in genetics and the increasing availability of health data have created an opportunity for personalized patient care to become a reality within the clinical field. Since the first human genome was sequenced in 2001 at a cost of around US$3 billion, the technology has become significantly easier and cheaper. Many genomes can now be sequenced within a day at a cost of approximately $1,000 each (Nature Outlook 2016). As a result, genome sequencing has entered medical practice as a procedure for diagnosing rare disorders where traditional diagnostic techniques have failed.

In contrast to a one-size-fits-all approach, in which disease treatment and prevention strategies are developed for the average person rather than by considering the differences between individuals, precision medicine provides patients with a unique approach based on genetics (National Cancer Institute 2017). The understanding of the genetic component of diseases has led to a new way of classification. Since all major diseases are characterized by a genetic component, their classification might be organized according to genetic differences in affected patients and not only according to their symptoms. Diseases can, therefore, be reclassified in molecular terms rather than by relying solely on gross pathology. Such a shift opens up opportunities to use the same drug or combination of drugs to treat diseases that are characterized by the same molecular basis. This difference would enable patients to be treated with more effective approaches that are targeted at their genetic profile.

1.2 Precision Medicine in Oncology

Although the novelty of precision medicine will inevitably permeate various medical branches, oncology—that is, the scientific study of cancer—is arguably in the vanguard. After heart-related diseases, cancer is the second leading cause of death in developed countries, and it was responsible for an estimated 9.6 million deaths in 2018 (WHO 2018). Globally, about one in six deaths is due to cancer. Although cancer can be partially prevented by avoiding risk factors and implementing existing evidence-based prevention strategies, oncology research aimed at developing effective treatments plays a key role in reducing cancer deaths. Many cancers have a high chance of being cured if diagnosed early and treated adequately. Research has shed light on different aspects of cancer that have made it possible to partially understand the biological mechanisms that cause cancer onset and its consequent development. This understanding is the cornerstone of the implementation of preventive measures and the development of cancer treatments.

The prevailing theory, which was formulated in the middle of the last century, interprets cancer as a set of about 200 diseases characterized by an abnormal cell growth that is released from the normal control mechanisms of the organism. All cancers begin in cells within human bodies. Cancer originates because changes in one cell or in a small group of cells determine an uncontrolled division. Proliferation (cell division) is a physiological process that takes place in almost all tissues and in countless circumstances. Normally, there is a balance between proliferation and programmed cell death (apoptosis), but mutations in DNA leading to cancer are responsible for the destruction of these ordered processes. This results in an uncontrolled cell division that may lead to the formation of a mass called a tumor (AIOM 2018).[4]

The place where a cancer starts in the body is referred to as the primary tumor or primary site. Eventually, cancer can spread to another part of the body. In order to spread, some cells must divide from the primary tumor and travel to another part of the body, where they start growing. This new area of cancer is called a secondary cancer or a metastasis. Some cancers may spread to more than one area of the body to form multiple secondaries or metastases. The more a cancer spreads, the harder in general it will be to eradicate (Cancer Research UK 2017).

The process leading to cancer—namely, the transformation of a normal cell into a neoplastic cell—occurs in various stages in which there is an accumulation of genetic, functional, and morphological aberrations. For this reason, the constantly varying molecular structure of tumors constitutes the research field in which the greatest hopes are invested for future clinical developments.

Within this framework, precision medicine will shed unique light on the study of cancer. One reason for this is that cancer is a genomic disease: most cancers harbor a cocktail of mutated (or altered) oncogenes[5] and tumor suppressors[6] that work in

---

[4]Most tumors follow this process, although not all: for example, tumors of the blood follow a different path.

[5]An oncogene is a gene that is a mutated (changed) form of a gene involved in normal cell growth. Oncogenes may cause the growth of cancer cells. Mutations in genes that become oncogenes can be inherited or they can result from being exposed to substances in the environment that cause cancer. See https://www.cancer.gov/publications/dictionaries/cancer-terms/def/oncogene.

[6]Tumor suppressor genes are normal genes that slow down cell division, repair DNA mistakes, or tell cells when to die (a process known as apoptosis or programmed cell death). When tumor suppressor genes do not work properly, cells can grow out of control, which can lead to cancer. See https://www.cancer.org/cancer/cancer-causes/genetics/genes-and-cancer/oncogenes-tumor-suppressor-genes.html.

concert to specify the molecular pathways that lead to their genesis and progression. To this end, oncology research has benefited greatly from the proliferation of worldwide efforts to characterize the genomes of thousands of cases spanning nearly all major cancer types. The same technological advances that have enabled a comprehensive catalogue of cancer genes are becoming increasingly applicable to advanced clinical diagnostics. Together with the expanding compendium of targeted anti-cancer agents in clinical development or active use, oncology has served as a proving ground for the genomics-driven framework that is unique among medical specialties (Garraway et al. 2013).

DNA alterations in the genesis of cancer stem from different causes, including environmental, genetic, and infectious aspects, causes linked to lifestyle, as well as random factors. Smoking, inadequate dietary habits, and inactivity significantly impact the risk of tumor development. Infections (like the papilloma virus) are responsible for an estimated 8% of tumors (AIOM 2019). Heredity can be a factor; for example, the genes *BRCA* 1 and 2 can increase the risk associated with the development of breast and ovarian cancer. However, heredity plays a low incidence in tumor genesis: less than 2% of the population carry mutations associated with hereditary neoplastic risk (AIOM 2019). Given that, in general, different risk factors are involved in cancer development, it is not easy to determine and assess the precise risk associated with single tumors; by definition, the genesis of a neoplastic disease is multifactorial. This means that an articulated combination of risk factors is concurrently at play in determining the onset and development of the disease. Hence, multiple factors should be analyzed within the patient's specific reaction, particularly in relation to the immune defense mechanisms and the processes to repair DNA damage (AIOM 2019).

When diagnosed with cancer, patients have typically received the same treatment as others who had the same type and stage of cancer, because tumors have traditionally been grouped and treated according to where they are found in the body. For example, all patients with a certain stage of kidney tumor would receive the same treatment. Nevertheless, different people respond differently to treatment, and, until recently, researchers could not understand the reason for this. After decades of research, scientists discovered that patients' tumors have genetic changes that cause cancer to grow and spread. They have also learned that the changes occurring in one person's cancer may not occur in others who have the same type of cancer, because tumors are often driven by unique combinations of DNA mutations. Furthermore, the same cancer-causing changes may be found in different types of cancer. Collectively, such changes are known as a tumor's mutation profile (Dana Farber Boston Children's Cancer and Blood Disorder Center [n.d.]).

When it comes to treating cancer, knowing which mutations are present in a tumor's profile may be more important than knowing where the tumor is located. The goal of precision medicine as applied to oncology is to treat patients with drugs that target the specific genetic mutations in their tumors, regardless of where the tumors are found. This extremely promising approach has already been implemented in clinical practice for certain kinds of tumors, and it is expected to become increasingly successful. Scientists are, therefore, working toward the development of genetic tests that will help decide the treatments to which a tumor is most likely to respond, which will protect patients from the unnecessary harm and stress caused by treatments that are not likely to work for them.[7]

In short, in the traditional approach to treating cancer, the treatment a patient receives usually depends on the type of cancer, its size, and whether it has spread; precision medicine, on the other hand, uses information about genetic changes in a specific patient's tumor, and this information is then used to decide the treatment that will work best for that patient.[8] Each patient has individual characteristics that differentiate them from others. In principle, therefore, each patient requires a tailor-made therapy. This is why precision oncology is so promising. Drawing on biological and clinical data, precision oncology aims to identify with the utmost accuracy the characteristics of cancer that affect the individual patient and then to build the best treatment strategy for that patient. It is now acknowledged that we can no longer speak about a single tumor common across cancer patients; rather, there are multiple tumors corresponding to multiple patients. The disease develops and progresses differently in each person. Moreover, there is evidence that genetic heritage, unique to every individual, interacts with the environment in ways that are specific to each patient.

The sum of genetics and epigenetics[9] results in unique patient responses to the disease and the therapies used to treat it. All cancers are characterized by mutations in DNA. These mutations have the potential to drive cancer growth, they can influence how a patient might respond to medication, and they might explain why a patient has stopped responding to treatment despite initially responding positively. By analyzing the tumor's underlying molecular alteration, physicians can prescribe targeted therapies tailored to each patient's condition and with the potential to improve the chances of recovery. In addition to targeting tumors with greater accuracy and increasing survival rates, research suggests precision medicine could mitigate unnecessary treatments and reduce prescription errors. The results so far have been promising. In a growing number of cases, precision oncology is becoming a reality, and it has led to an increase in the number of drugs available to hit the molecular characteristics present in the individual cancer. On the other hand, there are unfortunately many cases for which it remains impossible to offer personalized therapy. The hope is that research will make this number decrease progressively, because "targeted therapies"

---

[7]For further literature on this, see: André et al. (2014), Bedard et al. (2016), Begley and Ellis (2012), Chapman et al. (2011), Delbridge et al. (2016), Gray et al. (2015), Johnson et al. (2014), Kris et al. (2014), Lawler et al. (2015), Le Tourneau et al. (2015), Meric-Bernstam et al. (2015), Sohal et al. (2016), Slamon et al. (2001), Yap et al. (2011).

[8]See also the National Cancer Institute website at https://www.cancer.gov/about-cancer/treatment/types/precision-medicine.

[9]DNA modifications that do not change the DNA sequence can affect gene activity. Chemical compounds that are added to single genes can regulate their activity; these modifications are known as epigenetic changes. The epigenome comprises all the chemical compounds that have been added to the entirety of one's DNA (genome) as a way to regulate the activity (expression) of all the genes within the genome. The chemical compounds of the epigenome are not part of the DNA sequence, but are on or attached to DNA (*epi* means "above" in Greek). Epigenetic modifications remain as cells divide and in some cases can be inherited through the generations. Environmental influences, such as a person's diet and exposure to pollutants, can also impact the epigenome.

are one of the most important instruments of precision oncology: the cure is no longer a choice based only on the site of tumor development; rather, it is one that relates to the tumor's molecular characteristics, which can be different from patient to patient.

Thus, cancer treatment is changing from broad spectrum care (an "equal for all" therapy) to individualized therapies based on the singular genetic and the epigenetic traits of different types of cancer and the discovery of molecules that "block" the mechanisms that feed their development. The result of this transformation will be (and partly already is) the development of drugs capable of attacking the mechanisms causing the disease. The evolution of knowledge of how cancer develops at the cellular, genomic, and biochemical level, combined with the research ability to develop drugs able to interact with these mechanisms, underpin precision oncology.

Precision medicine in oncology therefore consists of medical intervention based on the specific biological characteristics of a specific tumor in a specific patient. The use of omics technologies involves grouping patients according to their molecular landscape and providing them with treatment based on their personal characterization of the disease. Understanding a more finely grained molecular landscape of an individual patient's tumor will be pivotal in the delivery of effective and safe treatment approaches that are targeted to patients' specific genetic profiles.

One way to perform such a personalized approach in oncology is to study patients' tissue samples collected at the time of surgery or diagnostic biopsy, in order for the molecular landscape of the tumor to be analyzed to find the therapy that best matches the profile. Unfortunately, the tissue samples retrieved by biopsies present only a partial picture of the tumor's molecular composition. In other words, the sample is a static representation of that specific portion of the tumor at a specific time. The limitations of this approach are apparent when intra-tumor heterogeneity is discovered and understood. Tumor heterogeneity refers to the way in which tumor cells can show different morphological profiles, including gene expression, metabolism, proliferation, and metastatic potential (De Bruin et al. 2014; Marusyk and Polyak 2010; Wang et al. 2015). This phenomenon can be observed both between tumors (inter-tumor heterogeneity) and within tumors (intra-tumor heterogeneity); it is the latter that this discussion focuses on. Because of intra-tumor heterogeneity, tumors have been discovered to be in constant evolution at a genetic and morphological level (Gerlinger et al. 2012; Russo et al. 2016; Siravegna et al. 2015); as a result, the molecular composition within the same tumor may not be homogeneous.

The heterogeneity of tumors presents significant challenges to planning effective treatment strategies, because intra-tumor heterogeneity prevents tumors from univocally responding to a single target therapy; consequently, there might be more than one target to which therapies have to respond. When a single tissue biopsy is performed, the resulting sample of tumor represents only a partial and limited picture of the total tumor within a specific time frame. Thus, focusing on the study of a simple tumor sample may fail to illustrate tumor heterogeneity, the appreciation of which is key to a successful targeted therapy. The primary tumor may be different from its metastases, and this needs to be considered when discussing therapeutic options. Research on heterogeneity can lead to a better understanding of the causes and progression of the disease. In turn, improved knowledge of heterogeneity can guide the creation of more refined treatment strategies that yield higher efficacy at a clinical level.

1.3   Tissue Biopsies

A successful application of precision medicine to the field of oncology depends on efforts to grasp tumor heterogeneity in its full extent; in doing so, the targets to attack in treatment can be identified. These advances can be applied in various ways. The benefit for the specific patient is that they may get treatments targeted at his or her specific genetic profile. In addition, greater understanding about a specific patient's tissue might reveal important mechanisms that foster research and contribute to the production of generalizable knowledge that might benefit other patients.

To study tumor heterogeneity when the tumor has developed in metastases, researchers should have access both to the primary tumor and to all its derived metastases that may be composed of the primary tumor's sub-clones. However, the primary tumor composition might not be homogeneous, and it may not share the same molecular landscape with all its derived metastases. In such cases, it might become essential to reach every tumor site in order to establish the most appropriate treatment path.

Unfortunately, access both to a primary tumor and to all its derived metastases in living patients through traditional tissue biopsy is difficult. Although many advances in oncology have relied on this procedure, major barriers exist in terms of the feasibility of collecting tissue samples from multiple sites for clinical aims. The procedure becomes even more complicated when the purpose of collection involves research aims.

Multiple tissue biopsies present several complications relating to the feasibility of this invasive procedure in end-stage cancer patients whose metastases may be spread all over the body. Moreover, many of these patients may opt for palliative and non-invasive care and no longer wish to explore invasive options. Besides the complicated logistics and the poor outcome, this procedure can cause unacceptable additional stress, pain, and anxiety for the patient; moreover, even if performed, it would provide only small tissue amounts that would result in a suboptimal resource for research targeted at developing precision oncology (Diaz and Bardelli 2014).

1.4   Rapid Tissue Donation (RTD)

Biological material for research aims can be procured from living persons, for example, through tissue biopsies. An alternative option of retrieval, however, involves the dead: biological samples removed postmortem are an extremely valuable resource for research and offer even more valid research opportunities than biopsies (Alsop et al. 2016; Van der Linden et al. 2014).

In light of this, RTD, a technique for multiple tissue collection from the dead, can overcome barriers posed by tissue biopsies in living patients and contribute to research advances in oncology (Schabath et al. 2014).[10] RTD is an advancing oncology procedure that involves procuring "fresh" tissue within 2–6 h[11] following the death of a cancer patient (Lindell et al. 2006).[12] On account of the short window between time of death and the collection of tissue samples, RTD is also known as a "warm" tissue donation that aims to preserve the high quality of specimens (Boyle et al. 2020).

Standard tissue biopsies provide researchers with a limited amount of material that is often suboptimal for research needs. However, end-stage cancer patients have a large tumor burden at the time of death, composed of the primary tumor and all its derived metastases, and this material, if collected after death, can offer significant opportunities for research; for example, it can provide great insights into the monitoring of the effects of novel therapies. Moreover, tumor tissue obtained through RTD may be useful for identifying targets for therapeutic intervention; for example, molecular changes at the time of treatment failure can provide relevant information about drug resistance. It is particularly relevant to understand treatment failure and cases in which tumors acquire resistance to a specific drug or combination of drugs— that is, understanding why a tumor becomes resistant to a treatment to which it had initially responded.

From the clinical and scientific perspectives, RTD provides researchers with several benefits. First, it enables the collection of large portions of tissues in the advanced stage of disease, a time when traditional biopsies are not performed because such invasive procedures would expose patients to unnecessary pain and anxiety (Schabath et al. 2014). Collecting material at an advanced stage of disease allows for molecular studies of disease progression and hence a greater understanding of the high morbidity and mortality of many cancers. Moreover, unlike in the case of traditional biopsies, no portion of tumor tissue needs to remain within the body of the patient, so researchers can collect a greater quantity of tumor tissue samples, the derived metastases, and, if necessary, unaffected tissues that are essential for control purposes. The unparalleled quantity and quality of tumor tissue retrieved with RTD is particularly important as it enables researchers to study the heterogeneity of tumor sites by comparing tissues and their molecular composition. RTD may provide useful information for measuring the progression of metastases as well as for analyzing differences in responses to a specific drug or treatment under research.

---

[10]Rapid autopsy programs began in the late 1980s. They were mainly focused on degenerative diseases such Alzheimer's disease, multiple sclerosis, and cancer. In this book, RTD is considered in relation only to oncology (see Pentz et al. 2005).

[11]While a window of 2–6 h is indicated as the ideal time for retrieval in order to preserve the quality of collected tissues, successful RTD yielding informative tissue up to 26 h after death has been reported (Boyle et al. 2020).

[12]It is worth noting that RTD significantly differs from other forms of research donation, such as donation of the whole body. The latter, although of great research value, involves rapid degradation of tissue, a problem that RTD is able to overcome.

The amount of material available facilitates researchers' ability to establish patient-derived cell lines[13] and xenografts[14] from primary and metastatic sites, which are recognized as the most appropriate models for studying the response and resistance to treatments. Cell lines are permanently established cell culture – that is, they are cultivations of cells grown under controlled conditions, typically in a lab. Patient-derived cell lines are when these cells were taken from a patient prior to the cultivation. In oncology research, patient-derived cell lines are frequently used in order to grow the same tumor of the patient in the laboratory so that its composition can be studied and possible treatment options analyzed. Patient-derived cell lines are generally used in combination with xenografts. The latter involves transplanting human tumor cells, either under the skin or into the organ type in which the tumor originated, into immunocompromised mice that do not reject human cells (other animals can be used as well, but mice are the most common animals used within this research area). This technique allows researchers to recreate a living environment in which the tumor may grow in order to study its development and to understand which treatments may be effective.

Another research benefit offered by RTD is the possibility of collecting tissues from both the primary site and its metastases in order to compare the molecular landscape of these two sources and to study tumor evolution. The access to a large quantity of high-quality tumor tissues through RTD may allow researchers to discover important biological insights into the frequency and nature of secondary mutations associated with tumor progression, as well as into the mechanisms of resistance to treatment. The type of research required to advance precision oncology cannot be performed on tissues previously collected and stored using preservation systems such as frozen or paraffin-embedded specimens,[15] since these means of preservation alter the molecular composition of tissues and may limit further research opportunities. Using RTD, on the other hand, enables primary tumors, metastases, and healthy organs to be retrieved with minimal risk of deterioration, thereby enhancing the research opportunities.

In summary, tissue collection via RTD offers a highly effective means to investigate the biology of primary tumors and a broad range of metastatic sites. This technique provides researchers with an unprecedented quantity of high-quality tumor tissue that is essential for advancing oncology research. The quality and the quantity of tissue retrieved is not comparable to that offered by other methods of collection. Major advances in oncology have occurred through the use of tissue biopsies and associated procedures. However, the utility of these samples is limited because of the small quantity of tissue that can be collected from living patients undergoing biopsies, and because of tissue deterioration after collection from already deceased

---

[13]Cell culture is a cultivation of cells derived from a multicellular eukaryote and grown under controlled conditions, outside their natural environment. A cell line is a permanently established cell culture that will proliferate indefinitely given an appropriate fresh medium and space.

[14]A xenograft is a graft of tissue taken from a donor from one species and grafted into a recipient from another species.

[15]Embedding tissue into paraffin blocks supports the tissue structure and enables very thin sections to be cut and mounted onto slides for microscopic analysis.

patients. Material collected via RTD may overcome the limitation of single tissue biopsy in relation to tumor heterogeneity, which is the key to studying effective treatments. The full extent and consequences of tumor heterogeneity can be evaluated by deep sequencing and global analysis of genetic alterations from several areas of the primary tumor and the metastases, and by correlating with clinical outcome. In other words, the opportunities created by the collection of fresh tissues through RTD are unparalleled and hold out the promise of being the basis for significant advances in the field of oncology.

1.5 Research Benefits of Postmortem Vs. Antemortem Tumor Specimens

It has been argued that RTD opens up opportunities that are not otherwise available through other means. Tumor tissue obtained from living patients in biopsies or in surgical resections is often inadequate for research needs. It may happen that tumors, such as early-stage small tumors or lesions that seem clinically benign, are surgically removed without being preserved in a way that allows further research on the tissue. Moreover, advancing technologies allow clinicians to perform minimally invasive biopsies in order to reduce the harm to the patient. The result is that the size of biopsy specimen has been sensibly reduced by leaving only a tiny portion (if any) for research purposes. However, a great complication associated with tissue biopsies is that collected specimens may not picture the heterogeneity of the tumor and may not represent the most biologically aggressive region of the tumor. As previously discussed, single samples of tumor retrieved via biopsies fail to depict intra-tumor heterogeneity and evolution (Spunt et al. 2012).

Metastatic tissues collected postmortem may be unique because they enable the "evolution of molecular abnormalities at different metastatic sites" to be investigated in a way that is otherwise not possible (Spunt et al. 2012). Moreover, tumor tissue obtained through RTD may be useful for identifying targets for therapeutic intervention: molecular changes described at the time of treatment failure can provide relevant information about drug resistance or may be valuable as biomarkers of aggressive disease or "drivers" of tumor progression. Ideally, these biomarkers can provide sufficient sensitivity to predict treatment success or failure earlier than diagnostic imaging, which unfortunately is able to detect tumor recurrence only at a relatively late stage. Identification of such changes "provides an opportunity to develop molecularly targeted therapies (i.e., therapies designed to target a specific gene or protein)" to inhibit tumor progression (Spunt et al. 2012, p. 3003). The advent of precision medicine—that is, of molecularly targeted therapies—underlines the need to identify specific molecular defects in high-risk cancers.

Traditional biopsies do not offer the chance to compare primary and metastatic tumor because of the poor quality and scarce quantity of material retrieved and available for research once diagnosis purposes have ended. Against this backdrop,

RTD offers the opportunity to implement the "molecular genetic autopsy", which allows a comparison between primary site and metastatic tissues in order to detect the frequency and nature of secondary mutations associated with cancer progression, metastasis to specific atomic sites, and the development of chemotherapy resistance (Balak et al. 2006; Schmidt-Kittler et al. 2003). In relation to this, recent studies (Gow et al. 2009; Wu et al. 2010) have shown that genetic findings may differ between a primary tumor and its derived metastases; such a difference may explain the failure of targeted therapies selected by relying on diagnostic tumor biopsy. As a result, a greater quantity of high-quality tissues retrieved postmortem may contribute to a better understanding of the evolution of tumors and their mutations, which are critical factors in developing targeted therapies. Nevertheless, RTD insights are not limited to revealing tumor heterogeneity and evolution: RTD may also help in the discovery of certain genetic features that are not lost during disease progression, thereby further contributing to the development of specific target therapies. The quality and quantity of material collected through RTD includes both affected and unaffected tissues, whose comparison may provide valuable insights into the root causes of cancer (Spunt et al. 2012).

More generally, the value of postmortem tumor tissue collection is not limited to advances in therapeutics. RTD can also contribute to the wider field of oncology by affording insights into the unexpected effects of given treatments. Studies (Brieva-Ruiz et al. 2008; Reed et al. 2000) have revealed that new cancer-related syndromes and findings that are clinically relevant to oncology patients are still being reported on the basis of autopsy material. To this end, published data from multiple autopsies provide an opportunity to study radiographic and clinical correlations, the side effects of new therapies, tumor responses, and tumor heterogeneity (Haas et al. 2007; Shah et al. 2004).

The new landscape opened up by precision medicine and its application in the clinical context presents promising opportunities for scientific progress. The possibility of developing personalized treatments according to the genetic profile of the individual patient has important implications for the development of various branches of medicine, with remarkable potential in the oncology field. The application of precision medicine in oncology is particularly valuable, because cancer is the second leading cause of death in the world and its rampant spread poses a real challenge to contemporary medicine. The understanding of how genetic traits are relevant in cancer management is likely to enable a significant shift from broad spectrum care to treatment administered on the basis of the understanding of singular genetic profiles of different types of tumors and the discovery of molecules that "block" the mechanisms that feed their development. This approach, which amounts to a new model of cancer management, would entail patients being treated not solely on the basis of their symptoms and tumor location, but also by being classified into subpopulations that differ in their susceptibility to a particular disease or their response to a specific treatment. Advances in precision oncology will facilitate more effective treatments for patients, targeted at their specific profile and increasing the chances of cure and a better quality of life.

In order to realize these promising outcomes, high-quality research in the field is needed.[16] Therefore, the study of cancer tissues constitutes a fundamental resource for understanding the mechanisms of onset and development of the tumor itself; this knowledge can then be translated into effective treatments for patients. The use of tissue biopsies in research can yield remarkable insights into tumor traits. However, a more extensive and comprehensive availability of cancer tissues, from both the primary site and its metastases, would present scientists with the opportunity to increase the quality of their research by allowing them to delve more deeply into a tumor's mechanisms. The retrieval of a greater quantity of high-quality tissue from living patients is not feasible because multiple tissue biopsies targeted at primary and metastatic sites are accompanied by several complications arising from the invasiveness of this procedure in end-stage cancer patients. Moreover, this procedure would expose patients to unnecessary pain and stress in a delicate moment of their lives. Consequently, cancer tissue retrieval from deceased cancer patients promises to be a better basis for scientific advances. The opportunity opened up by RTD would allow the barriers posed by tissue biopsy to be overcome, and it would increase the quality of cancer research's output by advancing the molecular understanding of tumors.

The implementation of "omics" technologies in research is the cornerstone of advances in precision medicine and holds out unique opportunities for the development of treatment. At the same time, however, it raises difficult and important ethical issues that involve, first, an adequate appreciation of risks and benefits to individual patients and their families, and, second, an appropriate balance between the risks to patients and the benefits to society. It is, therefore, vital to explore the ethics of informed choice and the responsible use of personal information in relation to RTD in order to develop policies that can simultaneously safeguard patients' rights while advancing scientific and clinical research.

References

AIOM—Associazione Italiana di Oncologia Medica. 2018. I numeri del Cancro in Italia. https://www.aiom.it/wp-content/uploads/2018/10/2018_NumeriCancro-operatori.pdf.

AIOM—Associazione Italiana di Oncologia Medica. 2019. I numeri del cancro in Italia. https://www.aiom.it/wpcontent/uploads/2019/09/2019_Numeri_Cancro-operatori-web.pdf.

Alsop, K., et al. 2016. A community-based model of rapid autopsy in end-stage cancer patients. *Nature Biotechnology* 10 (34): 1010–1014.

André, F., et al. 2014. Comparative genomic hybridisation array and DNA sequencing to direct treatment of metastatic breast cancer: A multicentre, prospective trial (SAFIR01/UNICANCER). *Lancet Oncology* 15: 267–274.

Aronson, S.J., and H.L. Rehm. 2015. Building the foundation for genomics in precision medicine. *Nature* 526 (7573): 336–342.

---

[16]For an example of research in the field, see the Cancer Moonshot Biobank, started by the National Cancer Institute at https://www.biospecimens.cancer.gov/programs/cancermoonshot/bio bank/default.asp.

Balak, M.N., Y. Gong, G.J. Riely, et al. 2006. Novel D761Y and common secondary T790M mutations in epidermal growth factor receptor-mutant lung adenocarcinomas with acquired resistance
to kinase inhibitors. *Clinical Cancer Research* 12 (21): 6494–6501.

Bedard, P.L., A.M. Oza, B. Clarke, et al. 2016. Molecular profiling of advanced solid tumors at
Princess Margaret Cancer Centre and patient outcomes with genotype-matched clinical trials.
Clinical Cancer Research 22: PR03.

Begley, C.G., and L.M. Ellis. 2012. Drug development: Raise standards for preclinical cancer
research. *Nature* 483: 531–533.

Boyle, T.A., G.P. Quinn, M.B. Schabath, et al. 2020. A community-based lung cancer rapid tissue
donation protocol provides high-quality drug-resistant specimens for proteogenomic analyses.
Cancer Medicine 9: 225–237.

Brieva-Ruiz, L., M. Diaz-Hurtado, X. Matias-Guiu, D. Marquez-Medina, J. Tarragona, and F. Graus.
2008. Anti-Ri-associated paraneoplastic cerebellar degeneration and breast cancer: An autopsy
case study. *Clinical Neurology and Neurosurgery* 110 (10): 1044–1046.

Buettner, R., J. Wolf, and R.K. Thomas. 2013. Lessons learned from lung cancer genomics: The
emerging concept of individualized diagnostics and treatment. *Journal of Clinical Oncology* 31:
1858–1865.

Cancer Research UK. 2017. Stages of cancer. https://www.cancerresearchuk.org/about-cancer/
what-is-cancer/stages-of-cancer.

Chapman, P.B., A. Hauschild, C. Robert, et al. 2011. Improved survival with vemurafenib in
melanoma with BRAF V600E mutation. *New England Journal of Medicine* 364: 2507–2516.

Cooper, G.M. 2015. Parlez-vous VUS? *Genome Research* 25: 1423–1426.

Dana Farber Boston Children's, Cancer and Blood Disorders Center. n.d. Precision cancer medicine.
http://www.danafarberbostonchildrens.org/innovative-approaches/precision-medicine.aspx.

de Bruin, E.C., et al. 2014. Spatial and temporal diversity in genomic instability processes defines
lung cancer evolution. *Science* 346: 251–256.

Delaney, S.K., et al. 2016. Toward clinical genomics in everyday medicine: Perspectives and
recommendations. *Expert Rev Mol Diagn* 16: 521–532.

Delbridge, A.R., et al. 2016. Thirty years of BCL-2: Translating cell death discoveries into novel
cancer therapies. *Nature Reviews Cancer* 16: 99–109.

Denny, J.C., et al. 2013. Systematic comparison of phenome-wide association study of electronic
medical record data and genome-wide association study data. *Nature Biotechnology* 31: 1102–
1110.

Diaz, L.A., and A. Bardelli. 2014. Liquid biopsies: Genotyping circulating tumor DNA. *Journal of
Clinical Oncology* 32 (6): 579–586.

Dzau, V.J., and G.S. Ginsburg. 2016. Realizing the full potential of precision medicine in health
and health care. *JAMA* 316 (16): 1659–1660.

Gahl, W.A., et al. 2016. The NIH undiagnosed diseases program and network: Applications to
modern medicine. *Molecular Genetics and Metabolism* 117: 393–400.

Garraway, L.A. 2013. Genomics-driven oncology: Framework for an emerging paradigm. *Journal
of Clinical Oncology* 31: 1806–1814.

Garraway, L.A., and E.S. Lander. 2013. Lessons from the cancer genome. *Cell* 153: 17–37.

Garraway, L.A., J. Verweij, and K.V. Ballman. 2013. Precision oncology: An overview. *Journal of
Clinical Oncology* 31 (15): 1803–1805.

Gerlinger, M., et al. 2012. Intratumor heterogeneity and branched evolution revealed by multiregion
sequencing. *New England Journal of Medicine* 366: 883–892.

Gow, C.H., et al. 2009. Comparison of epidermal growth factor receptor mutations between primary
and corresponding metastatic tumors in tyrosine kinase inhibitor-naive non-small-cell lung cancer.
Annals of Oncology 20 (4): 696–702.

Gray, S.W., et al. 2015. Marketing of personalized cancer care on the Web: an analysis of Internet
websites. *Journal of the National Cancer Institute* 107 (5): djv030.

Green, R.C., et al. 2013. ACMG recommendations for reporting of incidental findings in clinical
exome and genome sequencing. *Genetics in Medicine* 15 (7): 565–574.

Haas, G.P., et al. 2007. Needle biopsies on autopsy prostates: Sensitivity of cancer detection based on true prevalence. *Journal of the National Cancer Institute* 99 (19): 1484–1489.

Hunter, J.E., et al. 2016. A standardized, evidence-based protocol to assess clinical actionability of genetic disorders associated with genomic variation. *Genetics in Medicine* 18: 1258–1268.

Jain, K.K. 2009. *Textbook of personalized medicine.* New York: Springer.

Jameson, J.L., and D.L. Longo. 2015. Precision medicine: Personalized, problematic, and promising. *New England Journal of Medicine* 372 (23): 2229–2234.

Johnson, G.L., et al. 2014. Molecular pathways: Adaptive kinome reprogramming in response to targeted inhibition of the BRAF-MEK-ERK pathway in cancer. *Clinical Cancer Research* 20: 2516–2522.

Khoury, M.J., and S. Galea. 2016. Will precision medicine improve population health? *JAMA* 316 (13): 1357–1358.

Khoury, M.J., et al. 2016. Precision public health for the era of precision medicine. *American Journal of Preventive Medicine* 50 (3): 398–401.

Kris, M.G., et al. 2014. Using multiplexed assays of oncogenic drivers in lung cancers to select targeted drugs. *JAMA* 311: 1998–2006.

Landrum, M.J., et al. 2016. ClinVar: Public archive of interpretations of clinically relevant variants. *Nucleic Acids Research* 44 (D1): D862–D868.

Lawler, et al. 2015. Changing the paradigm: Multistage multiarm randomized trials and stratified cancer medicine. *Oncologist* 20: 849–851.

Le Tourneau, et al. 2015. Molecularly targeted therapy based on tumour molecular profiling versus conventional therapy for advanced cancer (SHIVA): A multicentre, open-label, proof-of-concept, randomised, controlled phase 2 trial. Lancet Oncol 16:1324–1334.

Ledford, H. 2015. CRISPR, the disruptor. *Nature* 522: 20–24.

Lindell, K.O., et al. 2006. Lessons from our patients: Development of a warm autopsy program. *PLoS Medicine* 3 (7): e234.

Lolkema, M.P., et al. 2013. Ethical, legal, and counseling challenges surrounding the return of genetic results in oncology. *Journal of Clinical Oncology* 31: 1842–1848.

MacConaill, L.E. 2013. Existing and emerging technologies for tumor genomic profiling. *Journal of Clinical Oncology* 31: 1815–1824.

Marusyk, A., and K. Polyak. 2010. Tumor heterogeneity: Causes and consequences. *Biochimica et Biophysica Acta* 1805 (1): 105–117.

McCarthy, J.J., et al. 2013. Genomic medicine: A decade of successes, challenges, and opportunities. *Science Translational Medicine* 5 (189): 189sr4.

Meric-Bernstam, F., et al. 2013. Building a personalized medicine infrastructure at a major cancer center. *Journal of Clinical Oncology* 31: 1849–1857.

Meric-Bernstam, F., et al. 2015. Feasibility of large-scale genomic testing to facilitate enrollment onto genetically matched clinical trials. *Journal of Clinical Oncology* 33: 2753–2762.

Minikel, E.V., et al. 2016. Quantifying prion disease penetrance using large population control cohorts. *Science Translational Medicine* 8 (322): 322ra9.

Mirnezami, R., et al. 2012. Preparing for precision medicine. *New England Journal of Medicine* 366 (6): 489–491.

National Cancer Institute. Cancer Moonshot Biobank. https://www.biospecimens.cancer.gov/programs/cancermoonshot/biobank/default.asp.

National Cancer Institute. 2017. Precision medicine in cancer treatment. https://www.cancer.gov/about-cancer/treatment/types/precision-medicine.

National Research Council, Committee on A Framework for Developing a New Taxonomy of Disease. 2011. *Toward precision medicine: Building a knowledge network for biomedical research and a new taxonomy of disease.* National Academies Press.

Nature Outlook. 2016. Suppl. Precision Medicine, Vol. 537, No. 7619. http://www.nature.com/nature/journal/v537/n7619_supp/index.html.

Pentz, R.D., et al. 2005. Ethics guidelines for research with the recently dead. *Nature Medicine* 11 (11): 1145–1149.

Phillips, K.A., et al. 2017. Making genomic medicine evidence-based and patient-centered: A structured review and landscape analysis of comparative effectiveness research. *Genetics in Medicine* 19 (10): 1081–1091.

Reed, E., et al. 2000. Analysis of autopsy evaluations of ovarian cancer patients treated at the National Cancer Institute 1972–1988. *American Journal of Clinical Oncology* 23 (2): 107–116.

Richards, S., et al. 2015. Standards and guidelines for the interpretation of sequence variants: a joint consensus recommendation of the American College of Medical Genetics and Genomics and the Association for Molecular Pathology. *Genetics in Medicine* 17 (5): 405–424.

Ritter, D.I., et al. 2016. Somatic cancer variant curation and harmonization through consensus minimum variant level data. *Genome Medicine* 8 (1): 117.

Russo, M., et al. 2016. Tumor heterogeneity and lesion-specific response to targeted therapy in colorectal cancer. *Cancer Discovery* 6 (2): 147–153.

Schabath, M.B., et al. 2014. Health care providers' knowledge and attitudes about Rapid Tissue Donation (RTD): Phase one of establishing a Rapid Tissue Donation in thoracic oncology. *Journal of Medical Ethics* 40 (2): 139–142.

Schmidt-Kittler, O., et al. 2003. From latent disseminated cells to overt metastasis: genetic analysis of systemic breast cancer progression. *Proceedings of the National Academy of Sciences of the United States of America* 100 (13): 7737–7742.

Schughart, K., et al. 2013. Controlling complexity: The clinical relevance of mouse complex genetics. *European Journal of Human Genetics* 21: 1191–1196.

Shah, R.B., et al. 2004. Androgen-independent prostate cancer is a heterogeneous group of diseases: Lessons from a rapid autopsy program. *Cancer Research* 64 (24): 9209–9216.

Siravegna, G., et al. 2015. Clonal evolution and resistance to EGFR blockade in the blood of colorectal cancer patients. *Nature Medicine* 21: 795–801.

Slamon, D.J., et al. 2001. Use of chemotherapy plus a monoclonal antibody against HER2 for metastatic breast cancer that overexpresses HER2. *New England Journal of Medicine* 344: 783–792.

Sleijfer, S., et al. 2013. Designing transformative clinical trials in the cancer genome era. *Journal of Clinical Oncology* 31: 1834–1841.

Sohal, D.P.S., et al. 2016. Prospective clinical study of precision oncology in solid tumors. *Journal of the National Cancer Institute* 108 (3): djv332.

Spunt, L., et al. 2012. The clinical, research, and social value of autopsy after any cancer death: A perspective from the Children's Oncology Group Soft Tissue Sarcoma Committee. *Cancer* 118 (12): 3002–3009.

Starita, L.M., et al. 2015. Massively parallel functional analysis of BRCA1 RING domain variants. *Genetics* 200 (2): 413–422.

Van Allen, E.M., et al. 2013. Clinical analysis and interpretation of cancer genome data. *Journal of Clinical Oncology* 31 (15): 1825–1833.

van der Linden, A., et al. 2014. Post-mortem tissue biopsies obtained at minimally invasive autopsy: An RNA-quality analysis. *PLoS ONE* 9 (12).

Vargas, A.J., and C.C. Harris. 2016. Biomarker development in the precision medicine era: Lung cancer as a case study. *Nature Reviews Cancer* 16 (8): 525–537.

Wang, Y., et al. 2015. Detection of tumor-derived DNA in cerebrospinal fluid of patients with primary tumors of the brain and spinal cord. *Proceedings of the National Academy of Sciences of the United States of America* 112 (31): 9704–9709.

Willig, L.K., et al. 2015. Whole genome sequencing for identification of Mendelian disorders in critically ill infants: A retrospective analysis of diagnostic and clinical findings. *Lancet Respiratory Medicine* 3 (5): 377–387.

World Health Organization. 2018. Cancer: key facts. https://www.who.int/news-room/fact-sheets/detail/cancer.

Wu, J.M., et al. 2010. Intratumoral heterogeneity of HER-2 gene amplification and protein overexpression in breast cancer. *Human Pathology* 41 (6): 914–917.

Yamamoto, S., et al. 2014. A drosophila genetic resource of mutants to study mechanisms underlying human genetic diseases. *Cell* 159 (1): 200–214.

Yang, Y., et al. 2014. Molecular findings among patients referred for clinical whole-exome sequencing. *JAMA* 312 (18): 1870–1879.

Yap, T.A., et al. 2011. First-in-man clinical trial of the oral pan-AKT inhibitor MK-2206 in patients with advanced solid tumors. *Journal of Clinical Oncology* 29 (35): 4688–4695.

Chapter 2
Two Senses of Informed Consent

Abstract This chapter introduces the evolution of the relationship between patients and physicians as well as implications for informed consent. It then focuses on the two meanings of informed consent: the *substantial* sense and the *formal* sense. In doing so, it subjects the notion of informed consent to a philosophical analysis. The first sense of informed consent is a specific form of autonomous authorization; this meaning is separate from, but strictly bound to, the second sense, namely the institutional set of rules and policy governing this act. The advent of the *formal* meaning of informed consent can be traced back to the Salgo Case of 1957, which affected the long-established Hippocratic paradigm of the patient–physician relationship. This case thus marked the shift of the traditional duty to obtain consent to become a new, explicit duty to report information before obtaining consent. This development created the expression *informed consent*. The rise of informed consent as a foundation of clinical practice marked a significant step within the field as patients became empowered with the right of being informed about medical interventions over their bodies.

2.1 Rapid Tissue Donation (RTD) and Informed Consent

As discussed in the previous chapter, the advent of precision medicine has led to remarkable changes in how cancer is tackled. Instead of deciding therapies by associating patients with the type and stage of tumor, precision medicine groups patients according to the molecular profile of the tumor in order to choose the most appropriate targeted therapies. The role played by RTD in these changes is crucial, as it offers researchers the chance to collect tissues with the adequate characteristics to perform high-quality research.

Despite the promising role of RTD in the development of precision medicine, and thus within oncology research, its implementation within the clinical setting involves several issues relating to the informed consent process. First, there is a major question: Should RTD be governed by an informed consent process at all? In RTD, tissue retrieval takes place *after* death, a moment in which obtaining informed consent for interventions might be perceived as less obligatory (and irrelevant with

C. Mannelli, *The Ethics of Rapid Tissue Donation (RTD)*, The International Library of Bioethics 85, https://doi.org/10.1007/978-3-030-67201-0_2

respect to the subject of the intervention). Moreover, the quality and quantity of retrieved tissues are important for the development of science. Hence, in pursuit of a greater good and given that retrieval occurs after death, it might be argued, at least in principle and for the sake of a philosophical analysis, that there is no need for an informed consent procedure. On the other hand, even when the need for an informed consent to regulate RTD is recognized, critical issues come into play relating to the form of such a consent and to the ways in which such consent should be collected, so as to encourage a choice that reflects patients' wishes while at the same time avoiding barriers to the development of science aimed at the public good. The lack of consensus about an informed consent procedure to regulate this form of tissue collection clearly prevents its development and implementation within the clinical setting, creating barriers to cancer research and precluding cancer patients from the opportunity to donate tissues to research after death.

Given these quandaries in the clinical setting, an ethical perspective can provide the analysis with relevant insights. An ethical point of view will focus on the evolution of the relationship between patient and physician and the consequent implications for the notion of informed consent.

2.2   The Hippocratic Oath

A primary historical source for investigating physicians' responsibility within medical decision-making derives from the Hippocratic Oath, which appears to be the inspiration underlying the standard idea of Hippocratic thought. Developed to regulate professional responsibilities, the Hippocratic corpus was the first set of Western writings about medical professional conduct and duties, and it remains influential today (Faden and Beauchamp 1986, p. 61). In Hippocratic thought, the relationship between the physician and the patient is established according to a definite "set of beliefs about what was considered either a right or a wrong medical behavior", and it thus presents the physician as an "all-knowing authoritarian figure, who decides what is best for his patients" (Pellegrino 2006).

In this tradition, physicians, guided by their traditionally established authority and knowledge, act as guardians of patients' health and generally assume that there are objective criteria for evaluating what is best for the patient. Drawing on the Hippocratic Oath, the physician ensures that patients receive the therapy that best promotes their health and well-being in order to preserve and extend life as long as possible. To this end, physicians use their knowledge to evaluate the patient's medical condition and their stage in the disease process, and to select treatments most likely to restore the patient's health conditions or, at least, to ameliorate their pain. Within the Hippocratic paradigm, patients generally assent to physicians' evaluations based on an "objective good" aimed at protecting their lives. Accordingly, Hippocratic physicians adopt a "paternalistic approach" toward their patients by deciding what is medically appropriate for them.[1]

Not all commentators agree that the paternalistic duty to preserve patients' lives has classical origins. Darrel Amundsen has argued that the physician's duty to prolong life is not rooted in Hippocratic medicine. Although scholarly opinion differs as to how many (if any) of the treatises in the Hippocratic corpus were actually written by Hippocrates, there is a general scholarly consensus that Hippocrates was not the author of the Hippocratic Oath. The first known reference to the Oath was made by Scribonius Largus in the first century CE, but its actual composition date is unknown (Amundsen 1978). Claiming that the Oath is an esoteric document that is often incompatible with Greco-Roman medical practices, Amundsen argues, therefore, for a later origin of this document, concluding that the only duty common to all Greco-Roman physicians was "to help or at least to do no harm" (ibid.).

Regardless of its origins, the Hippocratic paradigm is built on a conception of medicine from which flows its specific duty. The Hippocratic tradition considers the body as capable of self-conservation and self-healing. Consequently, medicine is a therapy that is intended to support this process of self-preservation and return to the original condition of health. In other words, medicine is a therapy that aims to encourage the *vis medicatrix naturae*, namely the restoring power of nature. Medicine, as Leon Kass wrote, is "a cooperative rather than a transforming art", and the physician "is but an assistant to nature working within the body having its own powerful (even if not invincible) tendencies toward healing itself (e.g. wound healing and other regenerative activities, or the rejection of foreign bodies and the immune response)" (Kass 1985, pp. 232–233). Within this conception of medicine, the body is its own healer, while the physician is a "cooperative but subordinate partner who supplies the needed materials" (Kass 1985, p. 233). According to Kass, the physician described in the Oath has the duty to assist sick patients by assuming a paternalistic attitude. The physician is:

> not only one who seeks to prevent the ignorant and the self-indulgent from harming themselves. He has the knowledge needed to direct and inform the otherwise dangerously open and uninformed human appetites. It may at first seem strange to think that we human beings need such knowledgeable outsiders or that we do not know what is good for us, all the more so on the premise that our body is the primary healer and the doctor but the physician's assistant. But a body possessed of the power of reason and hence also of choice, is a body whose possessor may lead it astray, owing to ignorance or wayward impulse. The physician, the ally of our body and of those inner powers working toward our own good, supplies needed knowledge, advice, and exhortation. He seeks to keep us from self-harm and injustice. (Kass 1985, p. 233)

For the Hippocratic physician, therefore, the patient's explicit consent for an intervention is not at the core of medical practice because the ultimate, shared and univocal good to be pursued is already established and involves the preservation of life. Within the Hippocratic tradition, when someone is sick and calls a physician for

---

[1]For further literature on this, see: Brock and Wartman (1990); Burke (1980); Dworkin (1988); Eddy (1990); Gorovitz (1982); Ingelfinger (1980); Katz (1984); Macklin (1987); Marzuk (1985); Siegler (1985); Szasz and Hollender (1956); Veatch (1972, 1981).

help, consent is, in a certain sense, implicitly attached to the demand for assistance. A patient's request for the help of a doctor implies that there is an understood and tacit agreement about consent between the two. Traditionally, physicians did not have the duty to inform patients concerning their medical condition or to ask them for any consent to or refusal of intervention. As Kass claims:

> it is sick individuals, not society or mankind or some abstract idea, who are the beneficiaries of the physician's activity. Moreover, the sick qualify for his services because they are sick, not because they have claims, desires, wishes, demands, or rights. The healer works with and for those who need to be healed, those who are not whole. (Kass 1985, p. 233)

Within this perspective, consent to intervention and faith in the physician's opinion is included and implied in the patient's request for help from the physician, whose aim is solely and unambiguously to preserve human life. Moreover, the paternalistic physician does not need to justify the course of action to the patient, because the "therapeutic privilege" grants the physician the authorization to arbitrarily decide whether or not to inform the patient concerning diagnosis, prognosis, and required medical interventions.

2.3   The Evolution of the Hippocratic Paradigm

Regardless of the uncertain origins of the Hippocratic Oath, the Hippocratic corpus has profoundly shaped medical practice throughout history. In particular, it has guided the patient–physician encounter, and this aspect continues to be influential today. However, it is important to clarify this element of the tradition. The interpretation of the Hippocratic tradition has undergone consistent evolution and changes throughout its history. It would be absurd to suppose that the Hippocratic physician has remained unchanged over time. Thus, the Hippocratic tradition should be regarded as an abstract model: the original and ancient Hippocratic paradigm has resulted in a conceptual and theoretical reference model with specific traits and influences. One element of this model is the paternalistic attitude of the physician toward the patient, an attitude that has survived up to the present day.

The Cambridge Dictionary defines "paternalism" as "the practice of controlling especially employees or citizens in a way that is similar to that of a father controlling his children, by giving them what is beneficial but not allowing them responsibility or freedom of choice".[2] As Beauchamp and Childress (2013) clarify, this analogy with the father involves two different features of the paternal role: first, the father is guided by his understanding of beneficence, that is, by what he believes to be in his children's best interests; and secondly, the father makes choices on behalf of his children, not letting them decide for themselves. These two attitudes can easily be transposed to the medical setting: the physician may assume an authoritative position to determine the patient's best interests, which is warranted by superior training and knowledge.

---

[2]https://dictionary.cambridge.org/it/dizionario/inglese/paternalism.

The paternalistic model and its evolution over time have also been analyzed by Ezekiel and Linda Emanuel, who have focused on how different kinds of relationship between patient and physician may impact and shape the medical decision-making process. In their analysis, there are four models of this relationship, of which the paternalistic is the first (Emanuel and Emanuel 1992). They compared the paternalistic model to Weberian ideal types, namely abstract descriptions of patient–physician encounters deprived of the complicating details that are frequently referred to in the field. According to this analysis, in the paternalistic model, which the authors also term the parental or priestly model, the patient–physician interaction ensures that patients receive the interventions that physicians believe best promote their health and well-being. This model assumes that there are shared objective criteria for determining what is best, which means that physicians, with the tacit compliance of the patient, know what is in the patient's best interest. As a result:

> the patient will be thankful for decisions made by the physician even if he or she would not agree to them at the time. In the tension between the patient's autonomy and well-being, between choice and health, the paternalistic physician's main emphasis is toward the latter. (Emanuel and Emanuel 1992, p. 2221)

The relationship between patients and physicians, and the attitude of the latter toward the therapeutic options of the former, has important repercussions for the informed consent process. In the paternalistic model, the physician acts as the patient's "guardian, articulating and implementing what is best for the patient. As such, the physician has obligations, including that of placing the patient's interest above his or her own and soliciting the views of others when lacking adequate knowledge". Thus, in this model, the conception of "patient autonomy is patient assent, either at the time or later, to the physician's determinations of what is best" (Emanuel and Emanuel 1992, p. 2221). Although the paternalistic model can be extremely nuanced, it can be interpreted—at least in the abstract form of a Weberian ideal type as presented by Emanuel and Emanuel—as recognizing a limited autonomy of the patient within the decision-making process. It is the physician who is entitled to discern what is in the patient's best interests and to make decisions accordingly.

In addition to the paternalistic model, which is the first model described by Emanuel and Emanuel, the informative model of encounter is based on the physician providing the patient with all relevant information in order to decide which intervention to choose. The physician's role in this model is to explain information and then to become simply an executor of the patient's choices: the physician acts as a technical expert. The informative model assumes a clear distinction between facts and values. The patient's values "are well defined and known; what the patient lacks is facts. It is the physician's obligation to provide all the available facts, and the patient's values then determine what treatments are to be given". Furthermore, there is "no role for the physician's values, the physician's understanding of the patient's values, or his or her judgement of the worth of the patient's values" (Emanuel and Emanuel 1992, p. 2221). Unlike in the paternalistic model, the patient's autonomy in the informative model overrides the control of the physician; the latter's role is to provide information in a detached way and then to carry out the patient's choices.

The third model is the interpretative one, in which the relationship between physician and patient is grounded in a collaborative reconstruction of the patient's preferences and aspirations in order to make the most coherent choice. In the interpretative model, the physician functions as an advisor to the patient, and the patient's autonomy is underpinned by self-understanding.

In the deliberative model, which is the fourth model discussed by Emanuel and Emanuel, the physician acts like a friendly teacher evaluating the most suitable choice for the patient and engaging in moral discussion. Not only "does the physician indicate what the patient could do, but, knowing the patient and wishing what is best, the physician indicates what the patient should do, what decision regarding medical therapy would be admirable" (Emanuel and Emanuel 1992, p. 2221). In this model, the patient is empowered by the deliberative physician throughout the decision-making process.

Although these four models are intentionally emphasized and fail to reflect the nuanced and complicated nature of reality, they provide a clear and reliable abstract framework for theoretically understanding how the interaction between the physician and the patient may impact the decision-making process and informed consent in the clinical setting.

2.4   Two Senses of Informed Consent as a Framework

According to the analysis of Faden and Beauchamp (1986), the literature, policies, practices have attributed two separate senses and meanings to the notion of informed consent, an appreciation of which lies at the heart of a thorough analysis of the topic. These two entrenched, and yet starkly different, meanings capture the different connotations that have been associated with the term "informed consent" over time.

The first sense of informed consent is a specific form of autonomous authorization conferred by a patient or a subject. This meaning is separate from, but at the same time strictly bound to, the second sense of informed consent, namely the institutional set of rules and policy governing this act. In my proposed framework, I will consider the first sense of informed consent as a *substantial* sense, whereas the second sense of informed consent is the *formal* one.

The *substantial* sense of informed consent has been defined by Faden and Beauchamp (1986, p. 278) as "an autonomous action by a subject or a patient that authorizes a professional either to involve the subject in research or to initiate a medical plan for the patient (or both)". This authorization lies at the core of this sense of informed consent: in authorizing, patients both assume responsibility for what they have authorized and transfer it to another's authority in order for it to be implemented (Faden and Beauchamp 1986, p. 280). Patients must understand that they are assuming responsibility for what they have authorized, and that they are warranting another individual to proceed. This *substantial* sense of informed consent is concerned with the very act of authorizing, which needs to be autonomous for the individual. Therefore, this sense of the term reflects the authenticity and genuineness of what an individual personally feels to be an autonomous authorization regardless of the institutional and legal framework. It is precisely this focus on the autonomy of the act that has led me to consider this sense as a *substantial* one.

By contrast, the second sense of informed consent represents its *formal* meaning. This sense refers to a legally or institutionally effective authorization from a patient and describes a social practice that takes place within both the medical and the research settings. The effectiveness of this act is determined by its compliance with rules and requirements that define specific institutional practice in healthcare and research (ibid.). In this second sense, informed consent is law-oriented and reflects the conformity to the social rules of consent that require medical professionals to seek valid consent from patients before performing medical or research interventions. This public practice is regulated by specific norms that define when and if a certain consent or refusal provided by a patient can be considered valid. The second sense of informed consent is soaked in cultural and legal norms, and hence this sense is rooted in specific historical time. *Formal* requirements for informed consent focus on establishing the behavior of the consent-seeker and on determining procedures and rules for the context of consent. Such requirements and procedures are obviously more readily monitored and enforced by institutions. Currently, elements of informed consent that are believed to be essential for the validity of this practice are: a disclosure and proper communication of necessary information concerning a specific medical intervention; the patient's comprehension of this information; the voluntary choice of the patient; the legal status of the patient, namely whether the patient is considered legally able to make a sound decision; and the final consent or refusal of the patient (Faden and Beauchamp 1986, p. 274). This *formal* meaning of consent represents the other side of the essence of authorization: it concerns the institutional setting that creates the environment in which authorization can occur.

The *substantial* and *formal* senses of informed consent express different meanings. The *substantial* sense of informed consent is "analyzable as a particular kind of action by individual patients and subjects: an autonomous authorization" (Faden and Beauchamp 1986, p. 276). In contrast, the *formal* sense of informed consent is analyzable in terms of the "web of cultural and policy rules and requirements of consent that collectively form the social practice of informed consent in institutional contexts where *groups* of patients and subjects must be treated in accordance with rules, policies, and standard practices" (ibid.).

In this framework, the interpretation of informed consent relies on the double meaning of this concept. The *substantial* sense of informed consent, as autonomous authorization, is associated with a dimension that exists de facto. In this first sense, informed consent is an authorization that is not defined by tangible requirements, but rather one that derives from its *substantial* existence. Therefore, the *substantial* sense of informed consent lies underneath *formal* requirements, and, because of its intangible nature, it cannot be led back to any kind of established prerequisite. This idea of consent is associated with the dimension expressed by the Latin words *substantia* (essence) and *substare* (to lie under), as it lies underneath the space-time features of what is socially and legally established, and it is unaffected by the variability of these features.

In contrast, the *formal* meaning of informed consent implies a dimension that can be located within the requirements stipulated by the law. To this end, this sense has been defined *formal* because it aims to meet specific and established forms. By "form", I refer to the external aspect of an act, an aspect that is necessary in order for the social environment to become aware of it and for the legal order to confer relevance to it. In other words, the form can be considered as the means by which the *substantial* and essential expression of intention (the *substantial* sense of informed consent) acquires legal effect.

The relationship between the two senses of informed consent is deep: they are both separate and simultaneously tied together. Yet this relationship should not lead to any overlap. The *formal* and *substantial* meanings of informed consent, although entrenched, are deeply different, and it may happen that an informed consent given in the first sense does not represent a legally effective informed consent in the second sense. As Faden and Beauchamp (1986, p. 281) point out:

> because formal institutional rules such as federal regulations and hospital policies govern whether an act of authorizing is effective, a patient or subject can autonomously authorize an intervention, and so give an informed consent in sense 1, and not yet effectively authorize that intervention in sense 2.[3]

In order to consider the extent to which an informed consent provided substantially represents an informed consent that is formally valid, let us consider the example of two 17-year-old twins, Kate and Phoebe; due to their age, they are considered minors in their country. Phoebe was born with a serious kidney disease that led to the necessary explant of a kidney; consequently, she now has only one kidney. Nevertheless, this circumstance does not prevent her from living a normal life. However, they are suddenly involved in a serious car accident. While Kate is not significantly hurt, Phoebe unfortunately is: she has sustained severe damage to her sole kidney and she urgently needs a transplant. After detailed discussion with the team, Kate consents to donate a kidney to her sister. Since Phoebe's first kidney had been removed many years earlier, Kate had plenty of time to think about how she would have acted in circumstances such as these. As a result, the decision she makes reflects a well-balanced evaluation of the pros and cons.

This example reveals the difference between informed consent in its *substantial* and *formal* senses. Kate gave consent in its *substantial* sense, as an autonomous authorization. The consent she provided was autonomous: she received adequate information about risks and benefits, and her decision reflected a balanced evaluation that she had reached over a long period. However, Kate failed to provide a valid consent in its *formal* sense. Legally a minor, Kate is not entitled to give an informed consent that complies with the legal requirements. In other words, Kate gave a consent that was not effective according to the law and institutional policies, which in this case required that the person consenting be a legally authorized

---

[3]Sense 1 refers to what I have referred to as the *substantial* sense; sense 2 refers to what I have called the *formal* sense.

agent. Thus, in this example, although the minor[4] may autonomously authorize a medical intervention by providing consent in its *substantial* sense, the intervention is not effectively authorized in the *formal* sense of informed consent (Beauchamp and Childress 2013, p. 122). This illustrates that *substantial* informed consent may, in some circumstances, not fulfill the requirements of a *formal* informed consent. If Kate was living in a jurisdiction where she is not considered a minor (for example, where adulthood legally begins at the age of 16), both her *substantial* consent and her *formal* consent would be valid; but in the scenario described above (that is, in which Kate is considered a minor), it is only her *substantial* consent that is valid.

Just as there can be valid *substantial* informed consent and invalid *formal* informed consent, so too can there be valid *formal* consent and invalid *substantial* consent.

Policies governing the formal sense should be formulated to conform to the standards of the *substantial* sense; in other words, institutional requirements for *formal* informed consent should be intended to maximize the likelihood that the conditions of informed consent in its *substantial* sense are satisfied. The autonomy-based model of informed consent in its *substantial* sense should function as a standard for informed consent in the *formal* sense (Faden and Beauchamp 1986, p. 284). However, a significant discrepancy emerges in the relationship between these two senses of informed consent. Professionals obtaining consent under institutional criteria (that is, in the *formal* sense) may fail to obtain consent that conforms to the autonomy-based model (that is, in the *substantial* sense), and vice versa. Thus, an informed consent obtained according to its *substantial* sense may not be effective in its *formal* sense, and, *mutatis mutandis*, an informed consent conforming to the *formal* sense may not be effective in its *substantial* sense. Given these premises, a sound informed consent should be effective according to both the *formal* and the *substantial* senses.

The *substantial* sense of informed consent—that is, the active autonomous authorization given by the patient or the subject—reflects a *real* expression of autonomous authorization, regardless of institutional settings. In the example discussed above, the consent expressed by Kate entailed a spontaneous and autonomous authorization, given without concern for legal requirements and rules since she was a minor and thus legally not entitled to provide valid consent. In this first sense of informed consent, the standards to be met are those of an autonomous authorization that has nothing to do with legal entitlements. In this scenario, the *substantial* sense of informed consent expresses a *substantial* authorization, which takes place without the mediation of institutional settings.

On the other hand, the *formal* sense of informed consent is rooted in a society's legal context and code. This meaning thus represents a policy-oriented consent, which is determined by the enforcement of laws. Thus, this sense of informed consent expresses its *formal* meaning and the standards to be met are those of compliance with the legal requirements.

---

[4]Minors are not legally eligible to provide an "informed consent". They are asked to express their preferences through what is known as "assent", which needs to be associated with the informed consent of parents or legal guardians. Here, in order to keep the example clear, I refer to informed consent rather than to assent.

In light of the preceding discussion, my analysis of informed consent in relation to RTD is based on its two senses, which represent the *fil rouge* of my interpretation.

2.5   The Salgo Case and the Formal Sense of Informed Consent

The advent of what I have termed the *formal* meaning of informed consent was Justice Bray's decision in *Salgo v. Leland Stanford Jr. University Board of Trustees* in 1957. This case involved a patient named Martin Salgo who was paralyzed as a result of aortography and sued his physician on the grounds that he had not been informed about the existence of such a risk. The court's decision held that failure to disclose the risks and alternatives was cause for legal action on its own. More precisely, the court found that physicians were in charge of disclosing "any facts which are necessary to form the basis of an intelligent consent by the patient to proposed treatment".[5] In the Salgo Case, all the pertinent topics of consent—the nature, benefits, risks, and alternatives of a medical intervention—were conceived as information needed by patients to know what they were choosing. The case marked the importance of "information" as an essential aspect of clinical practice and as a legal requirement of "consent". As Faden and Beauchamp (1986, p. 125) have pointed out, the Salgo Case marked the evolution of the traditional duty to obtain consent into a new, explicit duty to "disclose certain forms of information and then to obtain consent. This development needed a new term; and so *informed* was tacked onto *consent*, creating the expression *informed consent*".

The Salgo Case marks the birth of informed consent in its *formal* sense, namely as a social practice according to which consent is considered to be "informed" if and only if it is obtained through procedures that satisfy the legal rules and requirements. The Salgo Case also had important implications for the patient–physician encounter, since its judgement obligates physicians to inform their patients about the "nature of their condition and its expected course, about the benefits and risks of any proposed treatment, and of alternative treatment or non-treatment. This new legal requirement was impressed upon physicians as a professional duty". Thus, the Salgo Case, in giving birth to the *formal* sense of informed consent, was a significant step toward the development of informed consent as we know it today.

The emerging demands of modern medicine, as well as movements for patients' rights, autonomy, and informed consent, have presented remarkable challenges to traditional professional ethics. The Hippocratic tradition was not adequately prepared to respond to this emerging dimension. The Salgo Case created a crack in the solidity of the long-established paternalistic role of physicians in medical decision-making by profoundly altering the patient–physician relationship.

---

[5] *Salgo v. Leland Stanford Jr. Univ. Bd. Trustees.* 154 Cal. App. 2d 560, 317 P.2d 170 (1957).

2.6 Informed Consent as a Core

The establishment of "informed consent" in its *formal* sense in the Salgo Case changed medical practice and the patient–physician relationship that had hitherto been governed by the Hippocratic tradition. The new conception of informed consent leans toward an empowerment of patients with the authority to consent to, or to refuse medical interventions proposed by physicians. At the heart of this empowerment is information, since it is only by being adequately informed that the patient is able (or has the power) to consent to (or to refuse) an intervention. The shift from the Hippocratic consent to informed consent is significant. Informed consent becomes the *foundation* of clinical practice (Mori 2011). As such, no medical intervention can be legally performed without having obtained the explicit consent of the patient, and to obtain that consent it is necessary to provide the patient with, among other things, the necessary information to be able to make a decision. Unlike in the Hippocratic framework, this new conception of informed consent is not enclosed in a request for medical assistance, except for in specific cases, such as in emergencies. Patients need to decide what ought to be done to their body according to their preferences and values. Crucially, the establishment of informed consent required that patients be properly informed in order for them to make decisions. Thus, the advent of informed consent involved a notable implication: the patient's supremacy over their body and life emerges from this shift of decisional power from the physician to the patient.

The advent of informed consent marks an empowerment of the patient, whose right to be informed and then to decide about what happens to their body in the medical setting became officially recognized. Informed consent as a foundation of clinical practice determines a change of paradigm concerning the patient's role within medical decision-making. Within the Hippocratic paradigm, the best interests of patients are understood, objectively and incontrovertibly, to be the preservation of their life by the Hippocratic physician. Thus, a key feature of the Hippocratic paradigm is that life should above all be safeguarded as the greater good. However, with the introduction of informed consent, patients emerge as moral agents able to make medical decisions about what is best for themselves. To this end, patients have the right to determine the course of action by themselves evaluating the good that should be prioritized.

As a result, the rise of informed consent as a foundation of clinical practice marked an important step in medicine by changing the dynamics of the patient–physician encounter, with significant implications for patient autonomy and the decision-making process. The Salgo Case, by opening up the passage from consent to informed consent, marked an important shift away from the Hippocratic paradigm and to a new conception of the patient–physician encounter that involves providing the patient with the information necessary for making medical decisions over their own body. This move away from the paternalistic conception of the patient–physician encounter essentially involves empowering patients. Moreover, the Salgo Case grounded this new understanding within the principle that a patient should provide a legally effective authorization for a particular treatment—that is, an informed consent in its *formal* sense. This *formal* sense of informed consent is bound, but at the same time separated from, the other sense of informed consent, namely the *substantial* sense that involves an autonomous action of the patient. The relevance of the two senses of informed consent represent the *fil rouge* of this book and will be discussed in relation to RTD in the next chapter.

References

Amundsen, D.W. 1978. The physician's obligation to prolong life: A medical duty without classical roots. *The Hastings Center Report* 8 (4): 23–30.

Beauchamp, T.L., and J.F. Childress. 2013. *Principles of biomedical ethics*, 7th ed. Oxford: Oxford University Press.

Brock, D.W., and S.A. Wartman. 1990. When competent patients make irrational choices. *New England Journal of Medicine* 322: 1595–1599.

Burke, G. 1980. Ethics and medical decision-making. *Primary Care* 7: 615–624.

Dworkin, G. 1988. *The theory and practice of Autonomy*. Cambridge: Cambridge University Press.

Eddy, D.M. 1990. Clinical decision making from theory to practice: Anatomy of a decision. *JAMA: The Journal of the American Medical Association* 263 (3): 441–443.

Emanuel, E.J., and L.L. Emanuel. 1992. Four models of the physician-patient relationship. *JAMA: The Journal of the American Medical Association* 267 (16): 2221–2226.

Faden, R., and T.L. Beauchamp. 1986. *A history and theory of Informed consent*. Oxford: Oxford University Press.

Gorovitz, S. 1982. *Doctors' dilemmas: Moral conflict and medical care*. Oxford: Oxford University Press.

Ingelfinger, F.J. 1980. Arrogance. *New England Journal of Medicine* 303 (26): 1507–1511.

Kass, L.R. 1985. *Toward a more natural science: Biology and human affairs*. New York, NY: Free Press.

Katz, J. 1984. *The silent world of doctor and patient*. New York, NY: Free Press.

Macklin, R. (1987). *Mortal choices*. New York, NY: Pantheon Books.

Marzuk, P.M. 1985. The right kind of paternalism. *New England Journal of Medicine* 313 (23): 1474–1476.

Mori, M. 2011. Una "analisi bioetica" dell'attuale disputa sul testamento biologico come estensione del consenso informato. *Notizie di Politeia* 27 (102): 53–80.

Pellegrino, E. 2006. Toward a reconstruction of medical morality. *American Journal of Bioethics* 6 (2): 65–71.

Salgo v. Leland Stanford Jr. Univ. Bd. Trustees. (1957). 154 Cal. App. 2d 560, 317 P.2d 170.

Siegler, M. 1985. The progression of medicine: From physician paternalism to patient autonomy to bureaucratic parsimony. *The Archives of Internal Medicine* 145 (4): 713–715.

Szasz, T.S., and M.H. Hollender. 1956. A contribution to the philosophy of medicine: The basic models of the doctor-patient relationship. *Archives of Internal Medicine* 97 (5): 585–592.

The Cambridge Dictionary Online version. https://dictionary.cambridge.org. Accessed Nov 2020.

Veatch, R.M. 1972. Models for ethical medicine in a revolutionary age. *Hastings Center Report* 2 (3): 5–7.

Veatch, R.M. 1981. *A theory of medical ethics*. New York, NY: Basic Books.

Chapter 3
An Informed Consent for RTD? the Ethics of Postmortem Procedures

Abstract This chapter broadens the discussion of informed consent by considering biomedical ethics more generally. First, it presents Beauchamp and Childress's analysis of the four principles of autonomy, beneficence, non-maleficence, and justice, which constitute the basis of biomedical ethics reflection. The framework offered by the principles is the starting point to navigate the issue of an informed consent to govern Rapid Tissue Donation (RTD), the analysis of which is the main focus of this chapter. Arguments against the need for any form of informed consent for postmortem procedures, such as those of Hillel Steiner, Aron Spital and Charles Erin, will be addressed first. Next, positions advocating a regime of conscription that requires no consent will be presented, with specific focus on the duty to make one's body available for the sake of science. This duty will be analyzed through the duty of *fairness* toward other individuals for having benefited from research progress and the duty of *beneficence*, since enabling research to be performed on one's body after death prevents harm to others and does not require serious sacrifice. Finally, the chapter considers arguments, such as those of Joel Feinberg and Ernest Partridge, that advocate, on different grounds, a regime of consent. The ultimate aim of this chapter is to offer a philosophical analysis of the debate surrounding the need for informed consent for RTD; to this end, this chapter intentionally leans toward abstract philosophy rather than concrete policy.

3.1 *Formal* Sense of Informed Consent for RTD

The previous chapter considered the centrality that informed consent has acquired over time, and how it has been central to the shift away from the paternalistic Hippocratic paradigm of the patient–physician relationship and toward the empowerment of the patient in medical decision-making. This empowerment is grounded in the patient's right to be provided with the information necessary for enabling an autonomous decision. In particular, the chapter discussed how the *formal* sense of informed consent, initially established in the US as a result of the Salgo Case and then extending to other countries via other cases, refers to a legally effective authorization from a patient. From its birth, informed consent in its *formal* sense has established itself with increasing strength in both the clinical and the research fields, and it is now at the heart of interventions in the medical field.

The centrality of informed consent provides useful insights for a discussion of the role that informed consent should play within Rapid Tissue Donation (RTD). However, since tissue retrieval in RTD takes place after death, and the specific features of the collected biological material offers opportunities for advances in oncology, the need for an informed consent to regulate this procedure may be questioned, at least for the sake of a philosophical analysis. The fact that retrieval happens after death and that retrieved material provides some sort of benefit for society have been raised by commentators as compelling reasons against the need for informed consent in postmortem retrieval of organs.[1] This argument might be applied—at the very least, as part of a philosophical analysis—to RTD as well. This position, which questions the need for an informed consent both in the *formal* and the *substantial* senses, is challenged by those who argue that informed consent is necessary to regulate postmortem procedures and thus RTD. In order to consider this argument fully, informed consent needs to be discussed within the broader context of biomedical ethics, and specifically in relation to four ethical principles, as presented in the next section.

3.2   The Four Principles as a Framework

Advances in biomedical technologies pose new challenges to the healthcare setting. Emerging issues, including not only informed consent but also patients' rights, public health responsibility, and research involving subjects, have challenged the existing tools of professional ethics and highlighted the need for an adequate ethical perspective from which to analyze and address quandaries.

A crucial contribution that helps to meet this need is Beauchamp and Childress's reference framework consisting of the principles of biomedical ethics. Beauchamp and Childress proposed a moral framework composed of four clusters of principles derived from common morality to approach biomedical ethics and to use as a guide when examining concrete issues in the medical setting. The four principles are autonomy, beneficence, non-maleficence, and justice, and they constitute the basis of biomedical ethics. The principles do not offer a precise guide or a one-size-fits-all solution to ethical issues. Rather, they function as an abstract, comprehensive, and general dimension whose specification, balance, evaluation, interpretation, and implementation facilitate the analysis of concrete cases. The essence of the principles, as well as their balance, weight, and prioritization, have been much debated.

---

[1] See Spital and Erin (2002). This argument will be extensively addressed later in this chapter.

Although widely adopted, they have met with strong criticism; nevertheless, they remain the main basis for analysis in medical ethics.[2]

The principles proposed by Beauchamp and Childress have important implications in various areas. For the purposes of this work, the focus will be on their implications in the research field, with a view to providing a framework for reflecting on informed consent for RTD. According to Beauchamp and Childress, the principle of respect for autonomy is understood as respect for the autonomous choices of individuals. Autonomy, as its derivation from the ancient Greek *autos* (self) and *nomos* (rule, law) indicates, means "self-rule that is free from both controlling interference by others and limitations that prevent meaningful choice, such as inadequate understanding". The autonomous individual "acts freely in accordance with a self-chosen plan" (Beauchamp and Childress 2013, p. 101). Respect for the autonomy of individual choices constitutes the "primary justification of rules, policies, and practices of informed consent" (Beauchamp and Childress 2013, p. 121).

Underpinning such autonomous choice are intentionality, understanding, and being in control of the choice, each of which can be regarded as a condition that needs to be satisfied in order for a decision to be autonomous. An act is intentional when it corresponds to the subject's "conception of the act in question" (Beauchamp and Childress 2013, p. 104). Furthermore, subjects must understand their act in order for it to be autonomous. Understanding can be hampered by several barriers in the clinical setting, such as illness, fear, psychological distress, and communication barriers. The third condition is that the act be free from external control. External control might take the form of other people's controlling influences on the patient (be they relatives, healthcare staff, third parties), which prevent the patient's act from being autonomous. Yet, such influences might also come from inside the patient; for example, mental illnesses may limit or impair the voluntariness of the act. An individual acts autonomously, therefore, when their act is intentional, when they act with understanding, and when they act without any external controlling influence.

The first condition for autonomous action is clear-cut, but the second and the third both involve a matter of degree. An act is either intentional or it is not, yet the extent of its understanding by the subject, as well as the extent to which it is free from controlling influences, can vary to a greater or lesser extent. Thus, there can be degrees of autonomy: the autonomy of a choice sits within a continuum on a gradient.

The principle of beneficence requires individuals to take positive steps to provide benefits to others, and to balance benefits, risks, and costs to achieve the best overall results (Beauchamp and Childress 2013, p. 202). Beauchamp and Childress (2013, pp. 204–206) distinguish between specific and general beneficence. Specific beneficence involves particular parties with whom the subject has a relationship or to whom the subject is committed, such as family. General beneficence, which is particularly relevant within the RTD framework, is directed to the general population; in other

---

[2] See also Beauchamp (1995); Clouser and Gert (1990); DeGrazia (1992); DeGrazia and Beauchamp (2001); Gillon (1994, 1995, 2003); Green (1990); Holm (1995); Jonsen (1991); Macklin (2003); Richardson (1990, 2000).

words, it concerns people with whom the subject has no specific personal relationship or ties. Most people would agree, at least in principle, on the obligations entailed by specific beneficence, but those related to general beneficence are more controversial. Most obligations descending from beneficence in the healthcare setting are related to entering into a profession with deontological and professional duties. Given this, the principle of beneficence has important implications in clinical research, as it requires a thorough evaluation of risks and benefits for research participants. This evaluation acquires greater complexity when it comes to balancing the benefits and risks for individuals whose preferences cannot be acknowledged. In such cases, the principle of beneficence involves the pursuit of the best interest standard, according to which the "highest probable net benefit among the available options, assigning different weights to interests the patient has in each option balanced against their inherent risks, burdens, or costs" (Beauchamp and Childress 2013, p. 228).

In the healthcare setting, the principle of beneficence can conflict with the principle of respect for autonomy, with implications for medical paternalism. Whereas physicians traditionally relied on their own judgement to evaluate information and the involvement of patients within the decision-making process, the acknowledgement of patients' right to information and involvement in modern medicine have raised the question of whether patients' autonomy should be prioritized over professional beneficence.

Frequently combined with the principle of beneficence is the principle of nonmaleficence. This principle finds its origins in the maxim *Primum non nocere* ("First do no harm") and it "obligates individuals to abstain from causing harm to others" (Beauchamp and Childress 2013, p. 150). In the clinical setting, the principle of non-maleficence and its commitment to refraining from harmful acts has important applications related to, among other things, negligence in standards of care, treatment refusal, under-protection of research subjects, and side effects.

The final principle is that of justice. As Beauchamp and Childress (2013, p. 249) highlight, the construction of a theory of justice that is able to embrace different uses of principles of justice has not yet been achieved. The concept of distributive justice refers to a "fair, equitable, and appropriate distribution of benefits and burdens" (Beauchamp and Childress 2013, p. 250), and it has important implications for the healthcare setting. Allocation of scarce resources in healthcare, triage, prioritization, and rationing are typical scenarios in which the principle of distributive justice is applied and challenged. This principle also has specific implications for research, notably concerning the equal distribution of the benefits and burdens of research, the exploitation of vulnerable populations, and discrimination.

The four principles briefly introduced above constitute an established framework for every discussion in biomedical ethics, and they function as a reliable approach to discussing the need for an informed consent process to govern RTD.

3.3 Is There a Need for Informed Consent to Govern RTD?

The consideration of the four principles of biomedical ethics, as well as that of the evolution of informed consent to become a central principle in the patient–physician relationship, provides a basis for reflecting on the role that informed consent should play within RTD.

Is there a need for informed consent to govern postmortem research? Are the wishes and choices (if any) of the dead in any way binding when it comes to carrying out research on their bodies? Can any argument be raised for the claim that the wishes of the dead should be honored and respected? These questions are at the heart of any discussion about the need for an informed consent, in its *formal* sense, for postmortem research and will be addressed here through the lens of abstract philosophy rather than concrete policy. To these questions, different answers have been provided. Some have argued that no consent is needed, either on legal grounds or in relation to autonomous authorization, in order for research or interventions to be performed on the dead (and thus for tissues to be retrieved in RTD). This position is grounded, first, on the idea that the choices of the dead (if, indeed, one can speak of the dead as having choices) cannot be valued in the same way as those of the living, and, second, on the argument that retrieved biological material may offer benefits for society—in the case of RTD, benefits involve opportunities for advances in oncology research. However, others have argued that these procedures, although they happen after death, should necessarily be regulated by a specific informed consent process.

In the next chapter, I will present a detailed argument that it is necessary for RTD to be governed by informed consent. Prior to doing that, however, it is first worth considering different positions that have been advanced for and against the position that informed consent is required to govern procedures occurring after death. I will first consider positions arguing that the dead have no rights, and hence that a deceased individual has no moral claim over their own body; according to this argument, a body can be considered abandoned by its previously living owner at death and is thus left to the decision-making of others. If strictly interpreted, this argument rules out the need for (and, indeed, the relevance of) informed consent to govern postmortem procedures. I then turn to address positions advocating a regime of conscription for postmortem procedures such as organ donation; in other words, such arguments maintain that there is no need for consent. An important feature of such arguments is that they are based on various duties to others that override an individual's preferences. I introduce these duties in this chapter and then return to them again in the next chapter to explore more fully their relevance to RTD. Finally, I will discuss positions advocating a regime of consent, which prepares the way for my thorough argument in favor of such a regime in the next chapter.

3.4 Against the Need for Informed Consent to Govern RTD

3.4.1 Hillel Steiner: The Dead Have no Rights

That deceased individuals can claim no rights (including rights over their bodies) is argued by the Canadian philosopher, Hillel Steiner. Steiner begins from the assumption that only moral agents with a capacity for choice can be considered rights bearers. Since the deceased are no longer moral agents—because they no longer exist—it follows, according to Steiner, that they cannot claim rights. As a consequence, the bodies of deceased individuals, including their tissues and organs, are considered as abandoned and unowned, as *res nullius*. As such, these bodies enter the status of natural resources to which every living moral agent has an equal entitlement. Steiner's strong position thus states that deceased individuals do not qualify to have any kind of rights over their body. It follows from this premise that no informed consent should be sought in life for the collection of human samples after death.

Steiner's argument challenges both senses of informed consent. According to his perspective, the dead have no rights. No autonomous authorization is needed for research to be performed on abandoned bodies, and there is no institutional requirement to which this procedure needs to comply; as a result, neither a *formal* nor a *substantial* sense of informed consent is deemed necessary. Steiner's key assumption that the dead have no rights leads to a situation in which no legal basis for an informed consent can be recognized. Steiner calls into question the need for an informed consent to regulate postmortem research and concludes by strongly rejecting its role.

The lack of rights recognized for the dead is logically extended by Steiner to future agents, namely future generations. These categories are not entitled to rights because they cannot currently be considered moral agents. While deceased individuals do not qualify as moral agents because they no longer exist, future agents do not qualify because they have no element of contemporaneity with existing agents (Steiner 1991, 1994), and hence they share with the deceased the exclusion from the Steinerian right to equal liberty. Furthermore, children[3] (at least until they are old enough) do not qualify as moral agents, and, as a consequence, are not entitled to rights. Until children reach the status of moral agents, they are the property of their parents who act as their guardians (see also De Wijze et al. 2009).

Steiner envisions self-ownership—that is, rights over one's own body—as a corollary of the fundamental principle of equal liberty (Steiner 1974). According to Steiner, the right of self-ownership follows from the fact that moral agents inhabit human bodies (Steiner 1998). Since moral agents have the right to equal liberty, Steiner claims that "bodies must be owner occupied" (Steiner 1994, p. 232). For the purposes

---

[3] According to Steiner, this also applies to animals and mentally disabled individuals. This position has been met with significant criticism, to which Steiner has responded by arguing that to deny rights to these categories does not mean that we have no moral duties toward them.

of the discussion here, it is important to acknowledge that Steiner's analysis is, therefore, based on the assumption that the dead are not entitled to rights and that their bodies are in effect abandoned. Consequently, since the dead have no rights, and tissue retrieval for RTD happens after death, it seems that no informed consent has to be sought to perform postmortem research in either a *formal* or a *substantial* sense.

3.4.2 Spital and Erin

Steiner's position denies any right to deceased individuals; in his view, no informed consent (in either sense of the term) should be required for postmortem research. An intervention on deceased individuals without the need for a previous informed consent process has also been argued from other perspectives. Aron Spital and Charles Erin, for example, have argued that deceased individuals cannot be harmed by posthumous events and thus that they cannot be harmed by the use of their bodily material (Spital and Erin 2002). Spital and Erin apply this interesting principle in an attempt to solve the tragic shortage of human organs that causes the deaths of countless patients every year all around the world.[4] The authors propose what they describe as a "rarely discussed" solution, namely the conscription of organs from cadavers on the basis that the dead cannot be harmed by posthumous events. Conscription of organ donation is not central to my discussion here, but the arguments raised by Spital and Erin in favor of this system are relevant to the discussion of informed consent for RTD. Under the plan proposed by the authors, all usable organs would be "removed from recently deceased people and made available for transplantation". Consent would be "neither required nor requested and, with the possible exception of people with religious objections, opting-out would not be possible" (ibid., p. 612). Spital and Erin anticipate possible objections to their proposed conscription:

> No doubt this proposal will initially evoke shock, mockery and even outrage among those who believe that consent is an absolute requirement for cadaveric organ procurement. But the ethical basis for this widely held view is far from clear. Indeed, perhaps because the idea that consent is necessary has been so readily accepted as a given, few authors have seen the need to justify this point of view. We believe that this is a mistake. We will argue that careful consideration of the relevant issues will show that consent for cadaveric organ removal is not ethically required and that conscription is actually preferable. (ibid., p. 612)

Thus, Spital and Erin challenge the ethical basis of informed consent, and they claim that consent for postmortem organ retrieval should not be mandatory from an ethical point of view. As a result, informed consent is not "ethically required" for any form of postmortem intervention.

Spital and Erin outline several advantages of organ conscription. First, this system would boost the number of organs available, and would avoid vital resources being buried and thus wasted. As a result, it would result in many lives being saved that would otherwise have been lost due to the long wait for an organ donor. Secondly,

---

[4] See also Emson (2003).

conscription is less costly than other systems of organ procurement. Under conscription, it would no longer be necessary to "convince people to donate organs" or to "train requestors to obtain and document informed consent, no need for donor registries, no need for complex regulatory mechanisms that would be required to avoid abuses under plans for financial incentives" (ibid., p. 613). Moreover, conscription would avoid consultation with families to gather consent for organ transplant, which would spare the healthcare team and the families stress and discomfort. The final advantage of conscription is that it satisfies the ethical principle of distributive justice of benefits and burdens, because all people would "share the burden of providing organs after death and all would stand to benefit should the need arise" (ibid., p. 613).

A conscription policy, such as that proposed by Spital and Erin, amounts to a violation of the principle of respect for individual autonomy, since it deprives individuals of the right to consent to or refuse a specific procedure. Nevertheless, the authors claim that the principle of autonomy does not apply to cadavers since it "makes little sense to speak of autonomy of a dead person" (ibid., p. 613). In Spital and Erin's view, consent is ethically relevant when it safeguards the autonomy of individuals; but this relevance does not apply to cadavers, which have no autonomy and cannot be harmed (Jonsen 1988).

While pertaining specifically to postmortem retrieval of organs, the argument raised by Spital and Erin might be applied, for the sake of an ethical analysis, to informed consent for RTD. Adopting Spital and Erin's analysis, there is no need for an informed consent in its *formal* meaning to regulate postmortem procedures such as RTD because cadavers have no autonomy and cannot be harmed. The only possible concession allowed by the authors is the option to opt out for religious reasons, although it is unclear whether Spital and Erin allow this or not. They argue that the religious objection is a "sensitive and important issue", but they comment that "even the protection of religious interests is not absolute. For example, these interests are not sufficient to prevent compulsory autopsy. Are not the reasons for conscription of cadaveric organs at least as compelling as those for autopsy?" (Spital and Erin 2002, p. 614). Thus, it is not clear whether the right to opt out for religious reasons is definitively recognized. However, if Spital and Erin do admit a (significantly restricted) chance to opt out for religious reasons, then this seems to leave room for the acknowledgement of a certain (limited) possibility of choice recognized to individuals. Nevertheless, the claim that only religion can offer reasons strong enough to opt out from such an enforced system is debatable. Spital and Erin do not adequately explain why religion should be prioritized as a basis for opting out more than other personal preferences unconnected to religious faith.

3.5 Duty to Make One's Body Available for the Sake of Science: No Consent Requirement

Thus far, the analysis has focused on two specific positions: those of Steiner and of Spital and Erin. Both positions advocate a policy that does not recognize informed consent in either sense of the term for postmortem actions on cadavers. Steiner grounds his argument on the assumption that the dead have no rights, and thus no consent should be sought for postmortem activity. Spital and Erin argue that consent is vital for safeguarding individuals' autonomy. However, the dead can claim no autonomy, so it follows that there is no ethical or legal basis for consent.

Other commentators, however, have argued for a duty to make one's own body available for research (in life or after death) for the sake of science. This position is based on two fundamental duties: the duty of *fairness* toward other individuals for having benefited from research progress, and the duty of *beneficence*, since allowing research to be performed on one's body after death prevents harm to others and does not require a sacrifice. This position, if strongly interpreted, would lead to conscription (and hence to a position similar to that of Spital and Erin), the application of which would represent the lack of any consent requirement.

3.5.1 Duty of Fairness

The duty of fairness, due to having personally benefited from the achievements of science, is at the heart of a position that advocates a duty to make available one's body to research. Fairness implies that individuals who benefit from the participation in cooperative social schemes have a duty to reciprocally assume the burdens that a direct involvement in these cooperative social schemes implies (Van Assche et al. 2015). Caplan (1984) and Harris (2005) have argued for a general duty to participate in medical research that can be extended to a duty to allow research to be performed on one's remains after death.[5] The duty of fairness is put into effect by the duty to personally contribute to the maintenance of public goods and the duty to avoid free-riding.

3.5.2 Duty to Personally Contribute to Public Goods

The duty to personally contribute to the maintenance of public goods exemplifies the duty of fairness. Pure public goods have two defining features: one is "non-rivalry", meaning that a good can confer benefits without reducing the amount available for

---

[5] See also Harris (2003).

others. The other is "non-excludability", meaning that a good should not be provided unless it is possible for others also to enjoy it.

Biomedical knowledge is intended as a public good; thus, those who benefit from knowledge resulting from biomedical research should contribute to its increase and development (Schafer et al. 2009). This argument can be applied to participation in research for both living and deceased individuals (Van Assche et al. 2015). Biomedical knowledge, as a public good, is characterized by "non-rivalry", which entails that an individual can benefit from biomedical knowledge without diminishing the portion of the good reserved for other individuals; and by "non-excludability", which implies that every individual should have the option to benefit from this public good (Clark et al. 2003). Therefore, since most individuals benefit from the results of biomedical research—for example, through innovative drugs and medical assistance – they have a duty to personally contribute to this public good by directly participating in research. If this is the case, then this duty might include postmortem donation of tissues through RTD.

3.5.3   Duty to Avoid Free-Riding

A second argument put forward in support of a duty of fairness is the duty to avoid free-riding. Individuals who benefit from the results of biomedical research, but without contributing to that research themselves, are free-riding on those who actively support research advances (Evans 2004; Harris 2005; Orentlicher 2005). In economics, the free-rider problem occurs when those who benefit from resources, goods, or services do not pay for them, which results in an under-provision of these goods or services (Baumol 1952). In other words, free-riding involves taking advantage of other individuals' efforts by refusing to bear part of the burden. Thus, since it might be argued that every individual (in developed countries, at least) has benefited from the results of medicine, there should be a recognized moral duty to increase this knowledge. In this framework, one of the numerous ways to "carry part of the burden" and increase the knowledge could be to allow research to be performed postmortem.

3.5.4   Duty of Beneficence

The duty of beneficence is another key argument in favor of the existence of a duty to make one's own body available for research. The duty of beneficence requires individuals to act in ways that prevent harm and confer benefit, and, as discussed above, it can be categorized into specific beneficence and general beneficence (Beauchamp and Childress 2013). The latter of these categories is especially relevant here. Harris (2005) claims that individuals have a moral duty to participate in medical research because its goal is to alleviate patients' pain and to make new treatments available.

3.5.5 Duty of Easy Rescue

Closely connected to beneficence, the duty of easy rescue is often referred to within this framework. This duty, which was first analyzed by the medieval theologian Thomas Aquinas, entered the field of bioethics with the work of Peter Singer and Michael Slote (Van Assche et al. 2015). Singer (1972) introduced this duty within the famous argument concerning the child drowning in a pond: the ethicist claims that there is a moral duty to prevent harm without sacrificing anything of comparable moral significance.

This argument raises relevant issues to postmortem donation of tissues, as it might be argued that RTD amounts to a moral duty to prevent harm (to future cancer patients) that does not involve sacrificing anything of comparable moral significance (allowing postmortem retrieval of tissues).

3.6 Arguments for the Need for an Informed Consent to Govern RTD

Opposing the arguments against the need for an informed consent are those who claim that an informed consent procedure is necessary to regulate postmortem research and thus advocate recognition of both the *formal* and the *substantial* senses of informed consent. Some proponents of this position ground their arguments on the assumption that the dead can be (at least, to some extent) harmed by what happens after their death. Given that every dead person is also (by definition) a formerly living individual who had wishes and values that not only pertained to their life but may also pertain to the future after their death, failing to respect the wishes and the choices of the dead would impair a social institution, and thus it would also amount to a harm to the living who will be concerned about their own postmortem wishes.[6] Thus, the assumption that some acts can damage a dead person underlies a commonsense intuition that we are reluctant to abandon, namely, the attribution of a moral status to the dead. For example, Hamer and Rivlin (2003, p. 196) argue that the dead have "surviving interests which are capable of being harmed and these interests provide moral reasons against their being used". This position is in line with commentators who claim that the dead have "posthumous interests" that should be honored.

3.6.1 Joel Feinberg

Advocates of "posthumous interests" claim that individuals have interests that survive their death, which means that individuals can be harmed even when no longer alive,

---

[6]I discuss this idea, and my notion of the "once-alive", in more detail in Chapter 4.

should their interests be violated. For example, Joel Feinberg argues that, although death marks the end of a person, someone can still be "harmed" or have "interests" surviving their death. In his powerful argument about "unaffecting harms", Feinberg applies to the dead the concept of harm experienced by the living. He starts with the assumption that individuals can be harmed by circumstances of which they are not aware. To the question "Is it true that what we do not know cannot harm us?", Feinberg answers:

> If someone spreads a libelous description of me, without my knowledge, […] so that I am, still without my knowledge, an object of general scorn and mockery, I have been injured in virtue of the harm done my interest in a good reputation, even though I never learn what has happened. That is because I have an interest, so I believe, in having a good reputation as such […] and that interest can be seriously harmed without my ever learning of it. (Feinberg 1997, pp. 305–306)

Feinberg's passage captures Aristotle's idea that a dead man is popularly believed to be capable of "experiencing both good and ill fortune—honor and dishonor, and prosperity and the loss of it among his children and descendants generally—in exactly the same way as if he were alive but unaware or unobservant of what was happening" (Aristotle, *Nicomachean Ethics* 1.10, quoted by Feinberg 1997, pp. 306–307). Thus, according to Feinberg, knowledge is not a necessary condition for causing harm before death, so this assumption does not change in any relevant way after death. Feinberg argues that all interests belong to some person, and a person's surviving interests are the "ones we identify by naming *him*, the person whose interests they were" (Feinberg 1984, p. 83). When this person is dead, this circumstance does not prevent us from referring to their interests if "they are still capable of being blocked or fulfilled, just as we refer to his outstanding debts or claims, as if they are still capable of being paid" (Feinberg 1984, p. 83). Feinberg argues, therefore, for the existence of "posthumous interests" according to which the dead can be harmed even after death. To strengthen his position, he emphasizes that the interests harmed by events that occur after death are interests "of the living person who is no longer with us, not the interests of the decaying body he left behind" (Feinberg 1984, p. 89).

Feinberg presents a particularly persuasive and appealing argument in favor of "posthumous interests". Nevertheless, applying the notion of interest to the dead does raise concerns. The attribution of interests to the dead seems troublesome even when considering the definition of interest provided by Feinberg, according to which the concept of interest is based on the capacity to experience "awareness, expectation, belief, desire, aim, and purpose" (Feinberg 1974, p. 61). These capacities are not consistent with the dead and, given this definition, it seems to some extent incoherent to apply the capacity of having interests to them (Partridge, 1981). However, assuming that the concept of interest cannot apply to the dead does not necessarily exclude the option of taking into account their wishes to some extent.

3.6.2 Ernest Partridge

Ernest Partridge, who has also considered the interests and wishes of the dead, begins his analysis from the impossibility of recognizing "posthumous interests"; however, he considers that the wishes of the dead should somehow be respected. According to Partridge, the living show general expectations of having their interests respected and of the wishes of the dead being respected, perhaps largely because they want their own wishes to be respected after death. This attitude reflects a feeling of frustration prompted by the fear that our wishes will not be respected once we are no longer among the living. However, affirming that the living may have interests in respecting the will of the dead is not to ascribe posthumous interests to the dead. As Partridge claims, it is in the interest of the living "that they maintain the stable and just institutions that secured the wishes expressed by the deceased during their lifetimes" (Partridge 1981, p. 261). The future "posthumous interests" of the living are safeguarded by their resolution to respect the "posthumous interests" of the dead.

According to Partridge, the living respect the dead by contributing to the moral sense, by maintaining the moral community, and by supporting just and stable institutions. The living thus act to safeguard their own expectation that they can make plans of their own on this basis. If, however, they violate and disrespect

> the "quasi-interests" of the dead, they diminish their own living anticipations of favorably affecting the conditions of life beyond the time of their own lives, through their chosen disposition of their own possessions and through a keeping of promises made to them. (Partridge 1981, p. 261)

Partridge thus builds his thesis on the concept of posthumous respect, which may lead to a more comprehensive account of respect for the deceased. According to Partridge, even though the interests of a person do not survive that person's death, we "may nonetheless affirm that, in a community of moral personalities and just institutions, we are not only permitted to give the dead their due, we are morally required to do so" (Partridge 1981, p. 264).

3.7 Conclusion

After considering the four principles in medical ethics, this chapter has discussed various arguments for and against informed consent to regulate RTD. Since retrieval happens after death and since retrieved material provides some sort of benefit for society, some commentators have claimed that there is no need for informed consent to the postmortem retrieval of organs. At least in principle, this argument might also be applied to RTD.

Among those who argue against a need for informed consent, different positions have been considered, including Steiner's view that only moral agents with a capacity for choice can be considered rights bearers. Since the deceased are no longer moral agents, it follows, in this view, that they cannot claim rights. As a consequence, the bodies of deceased individuals, including their tissues and organs, are to be considered as abandoned and unowned, and thus any kind of postmortem research can be performed on them without the need for any authorization. Spital and Erin argued that deceased individuals cannot be harmed by posthumous events and thus they cannot be harmed by the posthumous use of their bodily material. In this view, therefore, no informed consent need govern this kind of research. Another line of argument involves acknowledging the principles in support of a duty to make bodily material available after death, which, if strongly interpreted, would lead to an absence of the need for consent.

Finally, positions advocating the need for an informed consent for RTD have been presented. According to Feinberg, individuals have interests that survive their death, which means that individuals can be harmed even when no longer alive, because there may be postmortem violation of their interests. Thus, posthumous interests should be respected just as the interests of the living are. On different grounds, Partridge claims for the need of an informed consent. Partridge argues that the interests of a person do not survive that person's death, but there nevertheless remains a moral obligation to "give the dead their due" (Partridge 1981, p. 264).

In order to extend and deepen the discussion of the question of informed consent for RTD, the next chapter introduces the concept of the "once-alive" as a way of thinking about the rights and interests of the dead in relation to postmortem biomedical intervention. This forms the starting point for my own positive answer to the question of whether RTD should be governed by a regime of informed consent.

References

Baumol, W. 1952. *Welfare economics and the theory of the state.* Cambridge, MA: Harvard University Press.

Beauchamp, T. 1995. Principlism and its alleged competitors. *Kennedy Institute of Ethics Journal* 5: 181–198.

Beauchamp, TL., and J.F. Childress. 2013. *Principles of biomedical ethics*, 7th ed. Oxford: Oxford University Press.

Caplan, A. 1984. Is there a duty to serve as a subject in biomedical research? *IRB: A Review of Human Subjects Research* 6 (5): 1–5.

Clark, C., et al. 2003. Internal and external influences on pro-environmental behavior: Participation in a green electricity program. *Journal of Environmental Psychology* 23 (3): 327–346.

Clouser, K.D., and B. Gert. 1990. A critique of principilism. *Journal of Medicine and Philosophy* 15 (2): 219–236.

De Wijze, S., et al. 2009. *Hillel Steiner and the anatomy of justice: Themes and challenges.* New York, NY: Taylor and Francis.

DeGrazia, D., and T.L. Beauchamp. 2001. Philosophical foundations and philosophical methods. In *Methods of bioethics*, ed. D. Sulmasy and J. Sugarman, 33–36. Washington, DC: Georgetown University Press.

DeGrazia, D. 1992. Moving forward in bioethical theory: theories, cases, and specified principlism. *Journal of Medicine and Philosophy* 17 (5): 511–539.

Emson, J.E. 2003. It is immoral to require consent for cadaver organ donation. *Journal of Medical Ethics* 29: 125–127.

Evans, M. 2004. Should patients be allowed to veto their participation in clinical research? *Journal of Medical Ethics* 30 (2): 198–203.

Feinberg, J. 1974. The rights of animals and unborn generations. In *Philosophy and environmental crisis*, ed. W. Blackstone, 43–68. Washington, DC: University of Georgia Press.

Feinberg, J. 1984. *The moral limits of the criminal law: Offense to others.* Oxford: Oxford University Press.

Feinberg, J. 1997. Harm and self-interest. In *Law, morality and society: Essays in honour of HLA Hart*, ed. P.M.S. Hacker and J. Raz, 284–308. Oxford: Clarendon Press.

Gillon, R. 1994. Medical ethics: Four principles plus attention to scope. *The BMJ* 309 (6948): 184–188.

Gillon, R. 1995. Defending the four principles approach to biomedical ethics. *Journal of Medical Ethics* 21 (6): 323–324.

Gillon, R. 2003. Four scenarios. *Journal of Medical Ethics* 29 (5): 267–268.

Green, R.M. 1990. Method in bioethics: A troubled assessment. *Journal of Medicine and Philosophy* 15 (2): 179–197.

Hamer, C.L., and M.M. Rivlin. 2003. A stronger policy of organ retrieval from cadaveric donors: Some ethical considerations. *Journal of Medical Ethics* 29 (3): 196–200.

Harris, J. 2003. Organ procurement: Dead interests, living needs. *Journal of Medical Ethics* 29: 130–134.

Harris, J. 2005. Scientific research is a moral duty. *Journal of Medical Ethics* 31 (4): 242–248.

Holm, S. 1995. Not just autonomy: The principles of American biomedical ethics. *Journal of Medical Ethics* 21 (6): 332–338.

Jonsen, A.R. 1988. Transplantation of fetal tissue: An ethicist's viewpoint. *Clinical Research Journal* 36 (3): 215–219.

Jonsen, A.R. 1991. Casuistry as methodology in clinical ethics. *Theoretical Medicine and Bioethics* 12: 295–307.

Macklin, R. 2003. Applying the four principles. *Journal of Medical Ethics* 29 (5): 275–280.

Orentlicher, D. 2005. Making research a requirement of treatment: Why we should sometimes let doctors pressure patients to participate in research. *Hasting Center Report* 35 (5): 20–28.

Partridge, E. 1981. Posthumous interests and posthumous respect. *Ethics* 91 (2): 243–264.

Richardson, H. 1990. Specifying norms as a way to resolve concrete ethical problems. *Philosophy & Public Affairs* 19 (4): 279–310.

Richardson, H. 2000. Specifying, balancing, and interpreting bioethical principles. *Journal of Medicine and Philosophy* 25 (3): 285–307.

Schafer, O., et al. 2009. The obligation to participate in biomedical research. *JAMA* 302 (1): 67–72.

Singer, P. 1972. Famine, affluence, and morality. *Philosophy & Public Affairs* 1 (3): 229–243.

Spital, A., and C. Erin. 2002. Conscription of cadaveric organs for transplantation: Let's at least talk about it. *American Journal of Kidney Diseases* 39 (3): 611–615.

Steiner, H. 1974. The natural right to equal freedom. *Mind* 83 (330): 194–210.

Steiner, H. 1991. Markets and law: The case of environmental conservation. In *The market and the state*, ed. M. Moran and M. Wrights, 43–58. London: Palgrave Macmillan.

Steiner, H. 1994. *An essay on rights.* Oxford, UK: Blackwell.

Steiner, H. 1998. Freedom, rights, and equality. *International Journal of Philosophy Studies* 6: 128–137.

Van Assche, K., L. Capitaine, G. Pennings, and S. Sterckx. 2015. Governing the postmortem procurement of human body material for research. *Kennedy Institute of Ethics Journal* 25 (1): 67–88.

Chapter 4
An Informed Consent for RTD?
Honoring the Wishes of the Once-Alive

Abstract Following on from the discussion in the previous chapter about whether there is a need for an informed consent to Rapid Tissue Donation (RTD), this chapter claims that an informed consent procedure is necessary. Rejecting Steiner's arguments about the deceased status, the analysis employs the concept of "once-alive" to *express*, and not to forget, the past condition of life and to bridge it to the actual condition of death. According to this dimension, this chapter claims that there is a need for informed consent to regulate RTD, because (1) it is crucial to honor, even after death, the wishes (if any) that individuals have expressed during their lives, (2) when these preferences do not jeopardize other living individuals. A focus on the concrete benefits to society provided by RTD and on a prospective refusal of a cancer patient to donate tissues are examined through a comparison of RTD with organ donation by relying on the framework proposed by Spital and Erin. To conclude, the argument confutes the positions according to which the body should be made available to research on the basis of the duties of fairness and beneficence presented within the previous chapter.

4.1 Introduction

In order to answer the question of whether Rapid Tissue Donation (RTD) should be governed by an informed consent procedure, the previous chapter introduced some biomedical ethical principles as well as various arguments for and against the notion that the wishes of the dead concerning their body after death should be honored.[1] In this chapter, the analysis will be deepened through an argument that informed consent is a necessary requirement for RTD. The argument begins by introducing the concept of the "once-alive" as a new lens through which to view a cadaver, one

---

[1] I have used the term "honored" throughout because it precisely captures the double sense of respecting another person and their wishes, as well as keeping an agreement or fulfilling an obligation. Thus, we honor the dead by paying respect to the lives they led and the values they held, and we also honor agreements, contracts etc. In the case of a person's enrollment in RTD, this is (as I argue in this chapter) an agreement that should be honored in both senses of the term, except where to do so would put someone else at risk.

---

which involves a shift away from a binary approach that contrasts the living and the dead. This forms the basis for a proposition that there should be a conditional granting of the wishes of the "once-alive", and hence that those wishes should be ascertained through informed consent.

Having set out this principle, I then turn to a comparison between RTD and organ donation. This discussion is relevant for two reasons: first, there are important overlaps between RTD and organ donation (above all, both are procedures that occur after death) as well as significant differences; second, much of the biomedical ethical discussion about the rights and duties of the dead has been framed in relation to organ donation, as Spital and Erin raised. Thus, my analysis considers the proposition that RTD should be governed by informed consent by discussing various arguments relating to duties (fairness, avoidance of free-riding, beneficence, and easy rescue— each briefly introduced in the previous chapter) in relation to the specific nature of RTD.

4.2 The "Once-Alive"

To consider the dead only as abandoned cadavers belonging to the category of those who no longer exist, or, as Steiner suggests, as natural resources that every moral agent has equal entitlement to, seems limiting. The recently dead are bodies of individuals who have led a life—long or short—that was shaped by choices, values, and aspirations. They established relationships, carried out projects, and developed interests and preferences. Given this, it would be restricting to consider the dead as belonging only to the category of abandoned cadavers. Rather than merely "deceased", I propose to consider this group as the "once-alive" in order to express rather than to forget the past condition of life and to bridge it to their actual condition of death. In other words, "once-alive" is a term that expresses this former aliveness, together with all the experiences, values, and qualities of life that shaped the life of a person. By shifting away from viewing a cadaver as "dead"—with all the finality conveyed by that term—and to regarding it as "once-alive", a new perspective can be adopted. Significantly, such a perspective captures the common way in which friends and family regard the deceased: not simply as dead but as someone who was once alive and whose life should be remembered and honored (for example, through rituals such as a funeral or a memorial service). Within the framework of RTD, choices and preferences that have shaped the once-alive, and that made these lives what they were, may come into conflict with the collection of tissues after death for research aims.[2] These preferences, wishes and choices—which had formerly been so meaningful for the once-alive—should not be disrupted in the name of RTD. Thinking of a cadaver as once-alive is a reminder—to researchers, clinicians, and ethicists, among others—of the former aliveness of the deceased person, the importance of this former aliveness

---

[2]It is worth noting that the analysis of informed consent in this chapter specifically concerns RTD; it is not intended to apply to other contexts.

to those who knew the individual, and the limitations of regarding that person as simply "dead".

In light of these considerations, I claim that there is a need for informed consent to regulate RTD, because (1) it is crucial to honor, even after death, the wishes (if any) that individuals have expressed during their lives, on condition that (2) these preferences do not jeopardize other living individuals.

This claim is built on two different points (1) and (2), the analysis of which can be broken down into two different sections. Let's start with (1).

1. *It is crucial to honor, even after death, the wishes (if any) that individuals have expressed during their lives.*

The importance of an informed consent to regulate RTD is particularly relevant because RTD involves unexplored ethical issues. The use in research of biological material retrieved after death may collide with the values and preferences the recently deceased individuals—the once-alive—had when they were alive. Research may thus constitute a threat to privacy and to the moral integrity of research participants who may oppose retrieval for various reasons, such as the bodily material not being used in a way that conforms to the values and preferences of those from whom the material was originally retrieved. Following Dworkin, "critical interests"[3] may be violated in a framework in which no informed consent is acknowledged to regulate postmortem research. Critical interests are those shaping individuals' projects and choices that give meaning to their lives. It may be argued, however, that it is inappropriate to ascribe critical interests to the deceased. With this in mind, I propose to refer to this category in terms of "quasi-interests" of the once-alive, namely those interests the deceased had during life and that the society of the living ought to honor, provided these interests do not jeopardize other individuals. When individuals have critical interests according to which plans and choices have been developed, it is important for the individual to know that these directions, which are meaningful for every individual, are respected even after death, or at least not disrupted so long as they do not involve risks to other living individuals. As a consequence, individuals should have their bodily material used (or left unused) in such a way that, as far as possible, it does not critically go against their life history, values, choices, and preferences. This is not to suggest that it is never permissible to ignore individual plans and choices (whether expressed by the living, or previously expressed by the once-alive). For example, in the case of dangerous infectious diseases, forced treatment may be permissible when its omission would result in a dangerous threat to society. In other words, coerced intervention can be justified if there is a grave risk to third parties. However, this circumstance does not apply to postmortem tissue removal, as will be further discussed below.

As a result, it is important that the critical interests that have shaped a person's life are honored even after death—at which point they become quasi-interests—provided

---

[3]This should not be confused with "experiential interests", which are related to the pursuit of pleasurable experiences.

that they do not jeopardize other individuals' lives.[4] This is relevant because those values contributed to making the once-alive person who they were (Dworkin 1986, p. 8). Postmortem research on cadavers can involve aspects that clash with the values and choices that an individual had during life. In other words, the research may amount to a violation of the type of person that an individual chose to be in life. For example, bodily material could be utilized for reasons that are incompatible with the once-alive's values and beliefs, and careful consideration should be given to these values and beliefs. Indeed, the destination of donated tissues is not straightforward, as frequently it is not possible to know in advance the plethora of uses for tissues, not least because research is based on exploration: experiments and observations can prove scientists' hypotheses wrong and may take them along unexpected paths. Research is in constant progress, so it is difficult to predict how it will evolve and advance; consequently, it is complicated to foresee how material destined to research will be used. Therefore, the complexity of delineating the boundaries of the possible future uses of collected tissues may represent an additional concern for prospective donors, especially because some uses may conflict with the personal or religious values of the once-alive.

Moreover, individuals should have the right to refuse RTD procedures if they have personal, cultural, religious, or any concerns. This clash of values—on the one hand, those chosen by the individual, and, on the other, those constituting a postmortem research protocol— would, even after death, be a violation of the values of the once-alive. It would lead, as Julian Savulescu points out, to a sort of instrumentalization:

> During life or *after death*, individuals are entitled to their body to be used in a way that is in line with their values. Each mature person should be the author of his or her own life. Each person has values, plans, aspirations, and feelings about how that life should go. People have values which may collide with research goals […] To ask a person's permission to do something to that person is to involve her actively and to give her the opportunity to make the project a part of her plans. When we involve people in our projects without their consent we use them as a means to our own ends. (Savulescu 2002, pp. 648–649; my emphasis)

Deciding not to regulate research on postmortem tissues with an informed consent procedure would imply putting someone else in charge of making decisions on behalf of the individual whose tissues were being retrieved. In other words, it would involve accepting that the personal informed consent of the individual be replaced by an obligation, identical for everyone, that fails to take into account personal preferences developed during life in a circumstance where no major public interest is jeopardized (an aspect that is further addressed below). This would lead to a leveling of the choices of individuals by attaching no weight to the personal values and choices made by individuals during life.

Given these premises, it is important, even in the case of postmortem tissue donation, to encourage an autonomous and responsible choice by the patient who must direct their will according to the system of values they have decided to embrace throughout life.

---

[4]This condition will be discussed in (2). It should temporarily be taken for granted, for the sake of the argument.

Encouraging and promoting informed consent means giving individuals the opportunity to lead a life based on choices that reflect, as much as possible, those chosen values. Conceiving an informed consent procedure for postmortem research means not only recognizing that an individual has lived according to a system of values, but also respecting their preferences even after death. In light of this, there is a need for informed consent to regulate this procedure.[5]

2. *Preferences should be honored when they do not represent a risk for other living individuals.*

At the core of this analysis is the recognition of a need for an informed consent for postmortem tissue donation that considers the dead not as mere cadavers but as those who were once-alive, with all the preferences and wishes living persons had. As such, their expressed wishes (if any) concerning the option to donate cancer tissues to oncology research should be honored and respected, but *with one condition*: these preferences and wishes should be honored provided that they do not represent a risk to other living individuals.

Point (2) in the claim for an informed consent procedure is, therefore, a condition to point (1) within the framework of RTD. An example of a preference of the once-alive that jeopardizes the living is a patient whom it is suspected has an extremely dangerous and contagious infectious disease at an initial stage. The patient gets accidentally hit by a car and dies. Just before her death, the patient made clear that she did not want clinical investigations or an autopsy on her body after death in order to ascertain if she had been infected with the virus. Honoring this wish would jeopardize other living individuals who might have been infected by the patient.

If the non-jeopardy of the living is the condition that makes it possible to honor the once-alive's preferences concerning RTD—and, therefore, makes it possible for this practice to be regulated by an informed consent—then it becomes vital to assess whether a refusal by a cancer patient to collect tissues for RTD may jeopardize other individuals' lives. If refusing the collection of tissues for RTD does in fact jeopardize other individuals, then the preferences of the once-alive, regulated by an informed consent, could not be honored. If, on the other hand, it can be argued that opposition to tissue retrieval for RTD does not jeopardize the living, then the preferences of the once-alive through an informed consent could be honored.

Before delving further into the argument that non-participation in RTD does not constitute jeopardy to others, it is appropriate to outline again the framework within which this analysis takes place—that is, the boundaries of RTD's purposes. As extensively discussed in Chapter 1, recent technological advances in genomics and proteomics are making possible a personalized approach to the diagnosis and

---

[5]It could be argued that, according to this reasoning, newborns, very young children, and mentally impaired people with cancer could be subject to tissue retrieval without their consent. In fact, if what matters is to honor the choices of each individual, then infants, children, and mentally impaired people who lack decision-making capacity could be excluded from the need for consent. However, if the objective of RTD is to honor the choices of individuals, then a fortiori the retrieval should not be encouraged in patients where it is not possible to assess whether RTD may be in line with their personal interests, preference, and choices.

treatment of tumors. Tissues retrieved through RTD provide a way to investigate the tumor biology of the primary site and all its derived metastases in a way that is not feasible by any other means. Consequently, the scientific value and potential benefits of RTD are promising. On the face of it, therefore, refusal to participate in RTD represents potential harm to others—that is, it does not satisfy the non-jeopardy condition—since it prevents the retrieval of tissues for research that promises to benefit living individuals.

The claim being advanced here, however, is that, in the face of the refusal of one or more cancer patients to adhere to RTD, there is no risk to other living beings if we consider *the advances that the tissue of that patient could have produced if used for primary purposes for which RTD is established*, such as improved understanding of tumor heterogeneity and tumor evolution. It is not being argued that a refusal to donate any kind of tissues (either cancer tissues or other kinds of tissues), which could possibly be used in an infinite array of purposes, might not also jeopardize the living.

The proposition that preferences concerning RTD should be honored because a cancer patient's refusal to donate would not represent a risk to other living individuals is based on two different arguments. The first (i) is that the quantity of tissues needed to carry out the primary purposes for which RTD is collected is limited; thus, when a proper RTD program is established, cancer patients who autonomously decide to enroll will meet researchers' needs without prospective refusal jeopardizing the living. The second (ii) argument concerns the concrete expectation of direct benefits for the living associated with RTD. The chance of direct benefits for the living associated with RTD is such that, in the face of a possible refusal, no life could be jeopardized. Each of these arguments will now be considered separately.

i. In light of these considerations, it clearly emerges that, within the framework of RTD, the primary aim for which tissues are collected is to observe specific aspects of tumors relevant to oncology research. This is because RTD, given its innovative technique, offers unique chances to understand tumor tissues in ways that are not possible by other means. Therefore, the main reason why any refusal to donate tissues is not likely to jeopardize the living is that the quantity of tissues needed to carry out the primary aims for which RTD is used—and not, therefore, other possible and indefinite uses for which retrieved biological material could be successfully used—is limited. Thus, once a properly structured RTD program is established and cancer patients are informed about the option to donate tissues after death, those who autonomously decide to donate their tissues provide researchers with an amount of biological material that meets the research needs in order not to jeopardize the living who are awaiting new cancer treatment paths.

ii. To acknowledge that research represents the primary endpoint of tissues retrieved through RTD means to establish a realistic expectation of the effective benefits that RTD will bring to society. Research is a long and winding process: typically, it takes years of studies and trials for drugs to be marketed, assuming the path is a successful one. Despite the scientific relevance of RTD within oncology research, the link between patients' donation of tissues and the translation into concrete benefits for cancer patients is neither immediate nor certain. Research proceeds according to a rigorous scientific method that involves formulating hypotheses based on observation, experimental measurement-based testing of those hypotheses, and the refinement or elimination of the hypotheses in light of experimental findings. Even the most successful and reliable scientist in the world can end up with unexpected results, no matter how well structured and well conducted the research project. Research is, by nature, uncertain. Given this, although RTD potentially yields valuable insights for cancer research, its translation into direct benefits for the living is not guaranteed. Consequently, if a cancer patient refuses to donate, it cannot realistically be argued that this decision jeopardizes other individuals' lives. Despite the scientific importance of RTD, this technique (unlike, for example, organ donation) cannot promise immediate benefits to other individuals, so the unwillingness to donate cannot be perceived as leading to jeopardy of others. Thus, since the reluctance of a cancer patient to donate through RTD does not pose substantial harm to other individuals, this procedure should be regulated by an informed consent process and not be subject to conscription.[6]

Nevertheless, someone may challenge this argument that there is an absence of harm to other individuals deriving from the refusal of a cancer patient to donate tissues through RTD. For example, K is a cancer patient. His cancer cells have such precious biological characteristics[7] that, if they were collected through RTD, they would lead researchers to find a miraculous treatment for cancer. Given the incidence of oncology diseases in the general population, would K's unwillingness to donate the precious cells to RTD jeopardize other individuals with cancer and future generations of cancer patients? Undoubtedly, given the special features of K's cells, these unfortunate cancer groups would have benefited from the new advances stemming from K's cells.

However, while conceptually fascinating, this argument loses strength when it is contextualized within the RTD research field. As noted above, it usually takes many years—and a huge amount of money—for new drugs to be marketed. Tissues retrieved from an RTD donor follow an extremely complex path: once collected, they are studied by researchers and cell lines are established. Tissues are generally inoculated in xenopatients in order to observe treatment responses. This path may lead to extremely promising results, yet there is no guarantee of concrete achievements either immediately or in the distant future. Given the high quality and quantity of material RTD provides, the main area to which RTD offers insights is the study of tumor heterogeneity. Nevertheless, it is difficult to imagine a cancer cell retrieved from Patient K whose tumor heterogeneity could provide more valuable information than that of a cancer cell retrieved from, say, Patient B. A concrete difference could eventually be observed in the case of rare tumors, whose incidence is so low that,

---

[6]This argument is intended for RTD only. I am not arguing that it should be applied to all types of tissue donation with other aims and purposes.

[7]Such as in the case of Henrietta Lacks. See Skloot (2010).

if Patient K refused to donate, it could be difficult to find "another source", namely a Patient B to collect tissues from. It is true, however, that if the tumor in question is so rare that it is untraceable in other patients, the potential damage associated with Patient K's refusal to donate his tissues for research aims remains significantly limited.

Given these considerations, it can be concluded that a refusal to donate tissues that can be collected only through RTD (in virtue of the particular conditions this technique offers) is unlikely to represent harm to the living. This consideration is valid for, and limited to, the research purposes for which RTD is primarily intended, like tumor heterogeneity, evolution, and response to treatment. Collected tissues can be used for an indefinite set of purposes, but only RTD purposes are covered by the preceding analysis.

It is clear that, if for some unusual and remote reason, the biological features of the cancer cells of Patient K were crucial to understanding the biological mechanisms underlying another pathology—depression, for example—that afflicts as many lives as cancer does, the failure to donate cancer cells potentially causes harm to living individuals with depression. However, researching the mechanisms underlying another pathology (in this example, depression) is not a primary purpose of RTD. Therefore, a refusal to donate tissues through RTD—in virtue of the particular conditions this technique offers and in virtue of the particular purposes for which this technique is intended—cannot be said to amount to harm to the living.

4.3   RTD and Organ Donation: Benefits to Society

The key points discussed so far, representing the core of the argument, are closely related to other aspects that demand further analysis. In order to make the discussion clearer and more effective, I will analyze these points within a comparison between RTD and organ donation, a well-known procedure of postmortem donation that shares some aspects in common with RTD. The case of organ donation was used by Spital and Erin (2002) as part of their argument against the need for an informed consent to regulate this kind of donation, since the urgency of supplying organs to save lives makes "conscription actually favorable". Even if the arguments of Spital and Erin are powerful in the case of organ donation, they cannot be successfully applied to RTD. While the organ donation literature offers a relevant and useful background for considering how to discuss and communicate strategies with prospective donors and their families, it does not provide an adequate basis for informed consent requirements to govern RTD. In fact, organ donation and RTD programs have widely differing aims and scope, rendering any attempt to find overlap between them dangerous.

Before proceeding, it is important to clarify that the following analysis is in no way intended to be an argument in favor of conscription for organ donation. The comparison between RTD and organ donation is used only to support my argument about the need for an informed consent to govern RTD.

The main difference between the two procedures is their purpose: whereas collected organs serve to be transplanted into another body to save a life—an end that has a direct, immediate, and certain social benefit—RTD collects biological material for research purposes, whose concrete return in terms of benefit to society is neither immediate nor certain. This substantial difference has several implications concerning the different probabilities of achieving the aims for which biological material is initially collected, the demand for necessary material and its balance with supply, and the different uses of donated material.

Organ donors know that their organs are destined to replace other people's diseased or damaged organs in order to provide them with a second chance of life. From the clear aim of organ donation, concrete aspects follow. Advances in science and technology have made organ transplantation a highly successful procedure: although adverse events can prevent organs from being effectively transplanted—for example, organs may not be healthy or they might be damaged during retrieval or transportation—this is a rare occurrence that happens in only a small percentage of cases. Thus, the great majority of donated organs are successfully transplanted to offer recipients a second chance of life, which means immediate and direct benefit for the living.

Despite the undoubted scientific value of postmortem tissue donation, immediate benefit is not what is at stake. Research on tissues has great potential to advance science and medicine in the long term, but in the short term the benefits are neither immediate nor certain. As discussed above, research involves exploration of the unknown, and thus, by definition, its discoveries and results can be unknown or unexpected. Donated tissue, since it is part of a research process, is not likely, therefore, to yield immediate results. Given this, whereas a refusal to donate organs could be argued to jeopardize the lives of the living, this cannot be compared to the jeopardy—if any—caused by a refusal to donate tissues through RTD.

The direct and immediate benefit of organ donation has a significant impact on the demand for necessary material and how this demand is balanced with supply. Every day, patients die while waiting for a new organ. With regard to RTD, however, the balance of demand and supply is extremely different due to the abstractness of the benefits offered by donated tissues. Whereas every organ serves to save a single life, tissues donated from a single cancer patient provide researchers with a huge amount of material to be used. Given the high-quality and high-quantity tissues yielded by RTD, a consent to RTD by a single patient may meet the research needs of a whole research team. In other words, the urgency associated with organ supply cannot be applied to tissues retrieved for oncology research, because the two procedures have different concrete benefits for society. Given this, it can be argued that the refusal to donate tissues through RTD is not likely to represent a jeopardy to other living individuals.

As discussed above, organ donation and RTD have different aims, with the result that their uses require different focuses. Since organ donation is clearly intended to save other lives, there is no unexpected implication concerning the use for which this material is intended. On the other hand, the destination of postmortem tissue donation is less straightforward. While organs are unequivocally destined to replace other organs, in RTD the situation is totally different. Frequently, it is not possible to present ex ante to donors the plethora of uses for which their tissues may be used, because it is difficult to predict how research will evolve and progress.

That the use of tissues retrieved for RTD is not straightforward has an impact on the potential implications that might arise from the different uses. Research on tissues might involve a breach of privacy because their use might reveal private health-related information of the donor (and perhaps blood relatives); this is not a possibility in the case of organ donations. Moreover, retrieved tissues might provide a great contribution to the development of new treatments that might be commercialized and provide significant income for pharmaceutical companies; again, this outcome does not apply to organ donation.[8] These possibilities, which apply only to RTD, need to be addressed and regulated through adequate informed consent procedures. Consequently, RTD raises concerns that are not comparable to those of organ donation procedures.

This structural difference, according to which the intended ends of donated tissues are unpredictable, can collide with the values and preferences of donors. This aspect is closely connected to the analysis presented above.

It has been argued that there is a need for informed consent to regulate RTD, as (1) it is crucial to honor the wishes (if any) that individuals have expressed during their lives, (2) provided that these preferences do not jeopardize other living individuals. The main point raised in (1) is that the need for an informed consent to regulate RTD is grounded in considering the dead not as *res nullius* but as once-alive—thus, as people who had lives shaped by values, projects, and preferences that should be taken into consideration when it comes to using their tissues after death. However, there is one condition: once-alive preferences should be honored only when they do not jeopardize the living. However, RTD is aimed at research needs and as such requires a limited quantity of biological material. For this reason, any refusal to donate to RTD is not likely to jeopardize the living, as, once an RTD program is properly established, those who voluntarily decide to donate would meet researchers' needs. Moreover, in the context of research, direct benefits to society are not guaranteed. The analysis of a lack of jeopardy for the living has been analyzed by comparing RTD and organ donation. With regard to the different probability of achieving the aims for which biological materials were collected, it should be considered that each organ is intended to save a life with a high chance of success, whereas RTD is aimed at research advances with neither immediate nor certain benefits. Concerning the demand for material and its balance with supply, it should be noted that each organ benefits a life, and people die every day due to organ shortage, whereas a single RTD donor may provide tissues for a whole research team. Concerning the possible uses of biological material retrieved, whereas organs are used to replace other organs, tissues might be used in a plethora of areas, an aspect that might raise concerns for donors and make an informed consent even more urgent.

Thus far, arguments supporting the duty to donate have been addressed. In the following sections, these positions will be analyzed by considering the various ethical

---

[8]This issue will be further discussed in the final chapter.

duties that were introduced in the previous chapter and which are frequently cited in relation to positions for and against conscripted organ donation: the duty of fairness and to contribute to public goods; the duty to avoid free-riding; the duty of beneficence; and the duty of easy rescue. Each will be analyzed with a focus on RTD.

4.4 Concerning the Duty of Fairness and the Duty to Contribute to Public Goods

An argument has been presented for an informed consent to regulate RTD. However, it might be argued that there is, above all, a duty to make one's body available for research (in life or after death) for the sake of science. If strongly interpreted, this claim would lead to conscription. This duty can be grounded in two different fundamental duties: the duty of *fairness* toward other individuals for having benefited from research progress; and the duty of *beneficence*, because allowing research to be performed on one's body after death prevents harm to others and does not require serious sacrifice.

The fulfillment of the duty to personally contribute to public goods through mandatory postmortem retrieval of tissues is hard to substantiate (Van Assche et al. 2015). First, not all research produces concrete results, such as public goods available to everyone. Greater attention should be dedicated to a fair allocation of research funds in order to encourage only projects that show a solid scientific basis. Nevertheless, it should be noted that scientific research is by its nature an exploration of yet-to-be-discovered areas; even if rigorously conducted, it can happen that research fails to provide concrete or expected results. Thus, it is essential to fund high-quality research projects, yet this will not guarantee that the results of the research are always positive.

Another argument against the duty to personally contribute to the maintenance of public goods is the unequal distribution of results. Individuals in underdeveloped countries have limited access to public goods, and it may happen that these populations are the very ones who carry the burden of research but lack adequate access to its resulting benefits. This argument relates to the equal distribution of research benefits—that is, the burdens of research are borne by individuals, but the results belong to pharmaceutical industries that financially benefit from the research and free circulation of results. However, these arguments, despite their power, somewhat miss the point. They do not question the status of scientific knowledge as a public good. On the contrary, they stress problems with the way research is conducted and how research results are distributed.

A final line of critique challenges the duty to directly participate in research by disputing the duty to concretely maintain the public good through the direct involvement of one's body after death. After all, almost every individual contributes to the public good—for example, through taxation and health insurance—and these actions should be considered sufficient to constitute a fair contribution to the public good. However, it might be noted that a financial contribution is not sufficient on its own, because medical research cannot rely solely on this; rather, it requires direct contributions by participants (Chan and Harris 2009; Forsberg et al. 2014). From this assumption, it follows that there is a need to concretely contribute to the public good by, for instance, allowing the postmortem use of one's body. However, this duty needs to be conditional; for example, the duty of fairness means that those who donate tissues to research while in life should be exempt from making such a donation after death (Van Assche et al. 2015).

In light of these considerations, it follows that the duty of fairness, as exemplified in the duty to personally contribute to the public good, raises an important point. However, this is a conditional duty, so it does not amount to a valid argument to support the existence of a duty to make bodily material available for research after death.

4.5 Concerning the Duty to Avoid Free-Riding

As in the discussion about the duty to contribute to the public good, it should again be noted that part of the burden of advances in research is ordinarily borne by individuals through taxation and insurance policies, aspects that would reject a free-riding thesis (Van Assche et al. 2015). On this basis, Chan and Harris (2009) have argued that specific biomedical research cannot rely only on financial contributions, since it requires the direct participation of individuals who provide material for research. Consequently, Chan and Harris maintain that there is an existing moral duty to contribute—not only on a financial level—to research progress. Since a direct involvement in research could be harmful or dangerous, this duty could be fulfilled through postmortem donation of tissues. However, as argued above, within the framework of the duty to personally contribute to the maintenance of public goods, the duty to personally contribute to science by allowing research to be performed on one's body after death seems to be a *conditional* duty (Van Assche et al. 2015). Drawing on the duty of fairness, those who directly contributed during life to the progress of science and to the maintenance of the public good, such as by donating tissues to research, seem to have already fulfilled their duty, and thus should be exempt from postmortem research carried out on their own body.

In conclusion, the duty of fairness, supported by the duty to personally contribute to the maintenance of public goods and the duty to refrain from free-riding, involves a conditional duty that should not be interpreted as binding; hence, it does not amount to a principle strong enough to justify a waiver to consent applicable to RTD.

4.6 Concerning the Duty of Beneficence

Several concerns might be raised over this duty when applied to postmortem research, given the real potential of direct benefit that RTD brings to society. The duty of beneficence requires an individual to support other individuals in many different ways, yet there are several ways that are considerably more effective than allowing postmortem research on one's body (De Melo-Martin 2008, pp. 13–14). Donating tissues through RTD does not represent the only means to fight diseases (Van Assche et al. 2015); for example, given the strong connection between poverty and disease, fighting poverty would achieve a greater result than donating tissues for postmortem research (Woolf et al. 2007). Given this, the duty of beneficence applied to postmortem donation of one's body for research constitutes a conditional duty, because the link between RTD and its translation into effective treatment paths is not immediate. After all, individuals may have strong reasons to donate their body to research, but it is up to them to decide whether to fulfill their duty of beneficence through this precise action or through another one. Thus, although postmortem donation has great potential for scientific progress, it does not amount to a duty, and, on this basis, individuals should not be required to donate through a conscription system.

4.7 Concerning the Duty of Easy Rescue

The duty of easy rescue is a conditional duty (Van Assche et al. 2015) and should not be interpreted as an argument for conscription concerning RTD. The conditionality of this duty has been analyzed by Beauchamp and Childress, who comment that some circumstances do not allow for discretionary choice about the beneficiaries of our beneficence. They propose that a person X has a "prima facie obligation of beneficence, in the form of a duty of rescue" toward a person Y apart from close moral relationships, if and only if each of the following conditions is satisfied (Beauchamp and Childress 2013, pp. 206–207):

1. Y is at risk of significant loss or damage to life, health, or some other basic interest.
2. X's action is necessary (singly or in concert with others) to prevent this loss or damage.
3. X's action (singly or in concert with others) will probably prevent this loss or damage.
4. X's action would not present significant risks, costs, or burdens to X.
5. The benefit that Y can be expected to gain outweighs any harms, costs, or burdens that X is likely to incur.

Within this framework, the duty of easy rescue is frequently cited in support of postmortem organ donation for transplantation, including as justification for a conscription system (Faber 2006; Hester 2006; Snyder 2009). As Spital and Erin have argued, conscription does not allow informed consent, either in its *substantial* or in its *formal* sense. Although this argument has been extended to postmortem tissue donation for research, this extension presents concrete issues. According to the framework proposed by Beauchamp and Childress, the first two conditions seem not to present major concerns when applied to RTD. Patients in need of innovative therapies may be at risk of significant damage to life or health. Moreover, from a scientific point of view, postmortem donation of tissues has the potential for advances that may constitute the basis for the development of new treatments. Thus, the first two conditions work for RTD as well. The third condition poses quandaries, since the link between postmortem tissue donation and the concrete prevention of loss or damage is uncertain, as previously discussed. Postmortem organ donation for transplantation is concretely associated with saving the life of an individual. However, this condition is not applicable to RTD because the actual correlation between the donation and its translation into a concrete health benefit for another person is uncertain.

The extension of the duty of easy rescue to postmortem donation of tissues for research aims challenges the fourth condition as well, since it involves concerns relating to the level of easiness of the rescue. After all, donating tissues postmortem may not involve an "easy" rescue. On the contrary, this action may present costs and risks for the individual by challenging their personal or religious values. For example, an individual may object to tissue donation on religious grounds, because they believe that the body should be buried unharmed. Thus, for such people, the cost of postmortem tissue donation would be disproportionate if compared to the effective benefit expected from their donation, so the act of donation is not for them an easy rescue. However, a duty to donate bodies after death may also pose concerns to individuals who have no specific religious objections to donation. After all, this kind of research may involve a breach of privacy and circulation of personal data; moreover, tissues may be used in ways that are incompatible with the moral values of the individual. If we consider organ donation, the aim for which donated organs are intended is clear from the beginning. In contrast, in RTD it is difficult to state in advance the ends for which donated tissues will be used. Moreover, given the strict window of time for tissue retrieval in the RTD framework, a cancer patient might be reluctant to impose on relatives a painful separation from their body immediately after death, in what will already be a particularly delicate moment. In light of these arguments, the duty of easy rescue can only partially apply to RTD, and it does not amount to an argument strong enough to support an ethical justification for a lack of informed consent.

Thus far, several arguments have been presented that maintain there is a duty to make one's body available for the sake of science, and thus that there is no need for an informed consent to regulate RTD. These duties, namely the duty of fairness—articulated in the duty to avoid free-riding and the duty to personally contribute to the maintenance of public goods—and the duty of beneficence—associated with the duty of easy rescue—amount to arguments in support of a duty to make one's body available for the sake of science. These duties have been interpreted as a reason to enforce a conscription to postmortem donation that implies a waiver to consent. Under such a conscription, neither *substantial* nor *formal* informed consent are compelling.

However, it has been argued that the duties marshalled to support the duty to make one's body available for science are conditional duties that are not strong enough to be considered full duties when applied to postmortem research. As such, these duties fail to adequately support the duty to make one's body available for the sake of science, which is a duty that implies conscription and a waiver to consent. Hence, although these arguments strongly support tissue donation, they are not strong enough to justify a conscription and thus a waiver to consent.

As a result, the arguments in support of a duty to donate tissues for postmortem retrieval do not justify a waiver to consent and thus the lack of necessity of informed consent in its *formal* and *substantial* senses. Whereas individuals can have extremely good reasons to donate tissues after death, they cannot be required to do so. These individuals should be strongly encouraged to—and they should be provided with the information to support this encouragement—but it remains up to them to decide whether and under what circumstances they are willing to donate their tissues for research after death through RTD.

4.8 The Need for Informed Consent to Govern RTD

Human bodily material removed postmortem through RTD is a particularly valuable resource for research in the oncology field. The need for an informed consent to regulate this procedure has been challenged on various grounds, among which are that no informed consent should be sought as retrieval takes place after death, and that collected tissues provide scientists with precious material for research. However, this perspective seems limited and hard to justify when applied to RTD in oncology research. Thus, this chapter has argued that there is a need for an informed consent to regulate RTD, as (1) it is crucial to honor, even after death, the wishes (if any) individuals have expressed during their lives, (2) provided that these preferences are not likely to jeopardize other living individuals. Within this framework, the main point raised by (1) has to do with the lens through which the dead are observed. The argument for the necessity of an informed consent to regulate RTD is grounded in considering the dead not as *res nullius*, or as mere cadavers to whom every living being has an equal entitlement, but rather as the once-alive, and hence as people who had lives shaped by values, projects, choices, and preferences that should be taken into consideration even after death. Yet, there is a condition (2): the preferences of the once-alive should be honored only when they are not likely to jeopardize other living individuals.

Therefore, if the non-jeopardy of the living is the condition that makes it possible to honor the once-alive's preferences concerning RTD—and, therefore, that make it possible for this practice to be governed by an informed consent process—then the key issue is to assess whether a hypothetical refusal to participate in RTD by a cancer patient would be likely to jeopardize other individuals' lives.

In order to make the lack of jeopardy to the living associated with an informed consent process to regulate RTD more evident, RTD has been compared with organ donation, a postmortem donation procedure with a different aim and whose resulting direct benefit to society is undoubtedly certain. The main difference between the two procedures concerns the demand for biological material and its balance with supply, a consideration especially relevant to the discussion on whether conscription is necessary. Organ shortage is a notable issue, and one that results in numerous deaths; by contrast, oncology research related to RTD requires only a limited quantity of biological material, so a cancer patient who decides to donate tissues can provide material for an entire research team, meaning that even a few such donations would ensure that there is unlikely to be a shortage of tissue. Secondly, the benefits to society associated with the two procedures (and thus the prospective jeopardy for the living resulting from a hypothetical refusal) are different: whereas each organ serves to save a life, the benefits to society of RTD are neither immediate nor certain. This difference has important implications for the probability of succeeding in the aims to which biological material is initially collected: whereas organ donation is highly successful (except in rare cases, retrieved organs are successfully transplanted into other bodies), RTD is intended for research, the outcomes of which are not guaranteed. Moreover, there is another aspect to consider that relates to the possible uses of retrieved biological material: whereas organs are unquestionably used to replace other organs, tissues collected for RTD might be used in a plethora of different areas of research, an aspect that might raise concerns for donors and make an informed consent process even more necessary.

In light of this comparison, a cancer patient's refusal to participate in RTD is unlikely ever to jeopardize a living being. As a result, there is a need for an informed consent to govern RTD. The condition is, clearly, that an effective RTD program is conducted within the oncology field, in order for cancer patients to know that this option exists and that it represents a way to foster scientific advances for future patients.

Against this backdrop, the next chapter, which discusses common ethical concerns relating to informed consent for oncology clinical trials, serves as a starting point to analyze the ethical issues raised by RTD. In doing so, it aims to develop an informed consent for RTD in its *formal* sense, and hence a legally recognized path that cancer patients can rely on. At the same time, the requirements for an informed consent are modeled on an autonomous authorization of the patient, namely on the basis of an informed consent in its *substantial* sense.

References

Beauchamp, T.L., and J.F. Childress. 2013. *Principles of biomedical ethics*, 7th ed. Oxford: Oxford University Press.
Chan, S., and J. Harris. 2009. Free riders and pious sons: Why science research remains obligatory. *Bioethics* 23 (3): 161–171.
De Melo-Martin, I. 2008. Response to Rosamond Rhodes. *Newsletter on Philosophy and Medicine* 7 (2): 13–14.
Dworkin, R. 1986. Autonomy and the demented self. *Milbank Quarterly* 64: 4–16.

Faber, C. 2006. *Whose body is it anyway? Justice and the integrity of the person*. Oxford: Oxford University Press.

Forsberg, J., et al. 2014. Why participating in (certain) scientific research is a moral duty. *Journal of Medical Ethics* 40 (5): 325–328.

Hester, M. 2006. Why we must leave our organs to others. *American Journal of Bioethics* 6 (4): W23–W28.

Savulescu, J. 2002. For and against: No consent should be needed for using leftover body material for scientific purposes. *The BMJ* 325 (7365): 648–651.

Skloot, R. 2010. *The immortal life of Henrietta Lacks*. New York, NY: Crown.

Snyder, J. 2009. Easy rescues and organ transplantation. *HEC Forum* 21 (1): 27–53.

Spital, A., and C. Erin. 2002. Conscription of cadaveric organs for transplantation: Let's at least talk about it. *American Journal of Kidney Diseases* 39 (3): 611–615.

Van Assche, K., et al. 2015. Governing the postmortem procurement of human body material for research. *Kennedy Institute of Ethics Journal* 25 (1): 67–88.

Woolf, S., et al. 2007. Giving everyone the health of the educated: An examination of whether social change would save more lives than medical advances. *American Journal of Public Health* 97(4): 679–683.

Chapter 5
Informed Consent and Research

Abstract This chapter discusses the importance of informed consent for research. It opens by focusing on the difference between research and care, the acknowledgement and appreciation of which is crucial for the informed consent process. An overview of regulations concerning the ethical conduct of research involving humans will be presented. Among these regulations, the Nuremberg Code, the Declaration of Helsinki, and the Belmont Report will be considered. Information, comprehension, and voluntariness are the basic features of an informed consent process according to the Belmont Report. However, when implemented within the concrete research setting, these three aspects raise relevant ethical issues. The analysis then turns to the extent to which the requirements enlisted by a *formal* sense of informed consent for research can provide an adequate framework to reach an autonomous authorization, namely an informed consent in its *substantial* sense. The aim is to present those ethical concerns raised by informed consent for oncology research that find some application—despite a necessary contextualization—in Rapid Tissue Donation (RTD). The analysis is divided into three sections dedicated, respectively, to information, comprehension, and voluntariness. The first section on information involves issues relating to the quality and quantity of data provided to patients. The second is dedicated to comprehension and includes a discussion of different kinds of benefits that might be involved in research and their assessment by patients. The final part is dedicated to voluntariness and addresses issues such as manipulation and the influences that might compromise the informed consent process.

5.1 Difference Between Research and Care: Ethical Issues

The introduction of informed consent in its *formal* sense has had a remarkable effect on clinical care, and it also plays a key role in another branch of medicine: medical research (Emanuel et al. 2008). According to the Code of Federal Regulations, the term "research" refers to a "systematic investigation, including research development, testing and evaluation, designed to develop or contribute to generalizable knowledge" (US Department of Health and Human Services, Office for Human Research Protection [n.d.]). Clinical research is a specific phase of research that involves human subjects. Research within the clinical field involves studies in which individuals participate as patients or volunteers (US Department of Health and Human Services, US Food and Drug Administration 2018), with their agreement to enroll being indicated by the informed consent form. This document generally consists of an information sheet and a consent form, and it includes information concerning the research at stake so that the individual can make an informed decision about whether or not to enroll in that specific protocol.

It is important to note how clinical care and clinical research differ, because the distinction between the two lies at the heart of the ethical conduct of medical research. Clinical care aims to benefit patients through therapeutic interventions chosen ad hoc for their specific medical condition. Clinical research, on the other hand, involves a systematic investigation, including research, development, testing, and evaluation, designed to develop or contribute to generalizable knowledge (US Department of Health and Human Services, Office for Human Research Protection [n.d.]). In other words, clinical research proceeds with a scientific method to demonstrate a hypothesis, and its ultimate aim is to produce scientific information that can be translated into generalizable knowledge. Whereas clinical care aims to provide personalized care, the ultimate goal of clinical research is to produce generalizable knowledge.[1] It is extremely important that patients and medical staff understand and appreciate the significance of this difference. However, within the clinical setting the boundaries of this demarcation may become dangerously blurred, which can lead to a plethora of ethical dilemmas.

In order to consider these dilemmas in relation to Rapid Tissue Donation (RTD), I first discuss how informed consent in research has acquired importance over time, before considering the difference between clinical care and clinical research by contextualizing it within the oncology setting. This will provide a concrete scenario for considering the importance of the emerging quandaries.

5.2   Regulations and Codes for Research Involving Humans

Informed consent holds a central place in medical practice, and its role in research involving human subjects is particularly relevant. Just as informed consent is the foundation of clinical practice, collecting consent from research participants is a central requirement for clinical trials. This has not, however, always been the case. Controversies over human subjects' involvement in clinical research have long been debated, fueled by scandals in which research subjects were exploited, injured, violated or killed. Recent history teems with outrageous abuses of research participants carried out in the name of "science"; in these cases, the consent of participants was not, or was only partially, obtained. Examples include the Tuskegee Syphilis Experiment[2]

[1] With the advent of translational research, the boundaries between research and care are less defined. However, it remains important to understand the different aims that animate the two dimensions.
[2] For further information on this topic, see Jones (1993).

(commonly known as the Tuskegee Study) in which consent was manipulated and "extracted from subjects in such social deprivation that manipulation came effortlessly" (Faden and Beauchamp 1986, p. 165) and the Nazi experiments during the Second World War.

Although informed consent for research gained particular relevance after the Second World War, its history can be traced back to the 1890s when the Prussian government, after a public scandal involving experimentation on unsuspecting patients who were inoculated with the spirochete that causes syphilis, required consent for any further research with human subjects (Moreno 1999). Later, Walter Reed, conducting research on yellow fever in Cuba, developed a contract—similar to the modern consent form—for his volunteers to sign; the contract included information about the risks involved (ibid.). Public concern about this issue in Germany culminated in 1931 with the promulgation of guidelines that required clear explanations of innovative or experimental treatments (Howard-Jones 1982). Interestingly, this pre-war German code of ethics, which addressed human experimentation, was, in some ways, more extensive in its protections and principles than either the postwar Nuremberg Code or the Declaration of Helsinki (Berg et al. 2001; Grodin 1992), both of which are discussed below.

To address ethical abuses such as those perpetrated by the Nazis, various codes for responsible conduct in the medical field have been adopted since 1945. The Nuremberg Code, which was drafted in 1947 as a result of the Nuremberg trials, entails a set of ethical principles for research involving humans. It represents an important landmark in medical ethics, since it acknowledges informed consent as a central and crucial aspect of participation in research. The Nuremberg Code sets out ten principles to satisfy ethical and legal concepts in order to regulate medical experiments involving human subjects (Nuremberg Code 1947).

Another important code is the Declaration of Helsinki, which was originally adopted in 1964 and has subsequently been revised. It sets out the importance of respect for individuals and for their self-determination, it pays particular attention to the protection of vulnerable people involved in research, and it stresses the right to informed consent for volunteers who enroll in research. The Declaration emphasizes the importance of information concerning the right of refusal and withdrawal: participants in research must be given adequate information about the aims, methods, anticipated risks, benefits, and discomfort, among other things (including the results of the study after it has concluded). In order to ensure that consent is not unduly influenced or collected under duress, it is emphasized that any refusal by a patient to enroll or any decision to withdraw from the study must never adversely affect the patient–physician relationship. The Declaration also addresses informed consent in relation to research involving human tissue: "For medical research using identifiable human material or data, such as research on material or data contained in biobanks or similar repositories, physicians must seek informed consent for its collection, storage and/or reuse" (Declaration of Helsinki 1964). However, the Declaration acknowledges the possibility of "exceptional situations where consent would be impossible or impracticable to obtain for such research. In such situations the research may be done only after consideration and approval of a research ethics committee" (Declaration of Helsinki 1964).

It is worth noting that these important norms were attempts to regulate human involvement in research. However, they did not guarantee an immediate and thorough establishment of informed consent in research, and their implementation in the broad research field encountered many difficulties. For example, the Tuskegee Study continued undisturbed until 1972 even after the Nuremberg Code and the Declaration of Helsinki had been issued. Similarly, the Nazi medical experiments took place despite the earlier German regulations on human involvement in research. As happened with informed consent in clinical practice, the shared acknowledgement and implementation of informed consent in research did not immediately follow from the promulgation of norms and regulations. The codes and declarations were issued in response to public scandals and research misconduct. Yet these regulations, when contextualized within the research setting, leave unsolved a plethora of ethical dilemmas that require further discussion in order to develop *formal* requirements of informed consent modeled on a *substantial* sense of it.

The Nuremberg Code and the Declaration of Helsinki were the precursors to the Belmont Report, which was published in 1979 and extensively outlined the importance and role of informed consent within research. The Belmont Report considers the principles according to which informed consent for research in its *formal* sense should be governed. This document is considered one of the most important ethical guidelines for research involving human subjects, and it has been embraced worldwide. The Report offers a useful and broad framework for understanding the fundamental requirements of informed consent in research. The basic principles envisioned by the Report are information, comprehension, and voluntariness, which have been recognized and adopted by every regulation on the topic. According to the Report, in order to provide their consent, subjects should be given specific and complete *information* about the medical intervention they are to undergo; and they must show *comprehension* of such information. There is also a requirement that the consent be *voluntary*, that it be free from external pressures, coercion or undue influence. However, these requirements, which constitute the basis of every regulation on informed consent, provide general assumptions about what informed consent for research should concretely look like. Thus, when these requirements are contextualized in the clinical setting, their abstractness results in the emergence of ethical issues. Concrete implications relating to the implementation of these requirements within the informed consent process will be highlighted by considering the context of the oncology research setting, whose structure and relevance provides fertile ground for discussion and paves the way for an analysis of the informed consent process for RTD.

5.3   Oncology Research as a Framework for RTD

The innovative nature of RTD brings with it significant ethical concerns that have to be addressed in terms of the informed consent process. Although some ethical concerns raised by RTD are new, others resemble those of medical areas with which RTD shares common ground, and hence areas that should be taken as relevant reference points. Among them, the area of organ donations has been argued to share some aspects in common with RTD. The organ donation literature offers a reliable basis for the approach to discussion and communication strategies with prospective donors and their families, but it does not provide an adequate basis for informed consent requirements to govern RTD. Indeed, organ donation and RTD programs have extremely different aims and scope, which might make any overlap dangerous in terms of consent procedures.

On the other hand, oncology trials share with RTD some common ground, in particular when it comes to analyzing ethical concerns associated with informed consent procedures. Above all, both RTD and oncology research involve research, which is very different from clinical care. This is particularly relevant for both informed consent processes, because patients need to comprehend and appreciate the gap between research and care and its implications for their expectation of any benefits resulting from interventions.

At the same time, however, the similarity between these two areas is only partial, because oncology clinical trials aim to recruit living patients, and, despite being in the early phases of research, carry at least the opportunity of a concrete benefit for those enrolled. RTD, on the other hand, seeks the consent of a living cancer patient for a procedure that will happen after death. Thus, it involves in effect a deceased population, for whom no direct benefit can be contemplated. As will be further discussed, this aspect of RTD involves ethical concerns related to an accurate risk–benefit evaluation, which donors must weigh up before making a decision.

5.4   Informed Consent for Oncology Research
and the Belmont Report's Requirements

Oncology research raises significant ethical quandaries with regard to informed consent. These quandaries partly relate to the difference between research and care and to the specific structure of oncology clinical trials. This scenario is relevant for the analysis of informed consent and the two senses associated with the term. The structure and methodology of oncology clinical trials is significantly complex, and oncology research may involve risks for participants. These aspects provide fertile ground for ethical discussion. Given the features of oncology research, individuals who are evaluating an enrollment must—as in every other form of research—be thoroughly informed about the research in order to be able to decide whether or not to participate; in other words, they need to be thoroughly informed in order to express their consent. Collecting patients' informed consent for research raises several ethical aspects that are exacerbated by the singular structure of oncology clinical trials and by patients' appreciation of this structure.

Institutional requirements for *formal* informed consent should be intended to maximize the likelihood that the conditions of informed consent in its *substantial* sense—namely, autonomous authorization—will be met. Given this, the autonomy-based model of informed consent in its *substantial* sense should function as a standard for informed consent in its *formal* sense. Starting from these premises, I will analyze the extent to which the requirements of a *formal* sense of informed consent for oncology research can provide an adequate framework to realize autonomous authorization, that is, an informed consent in its *substantial* sense. By drawing on the structure and relevance of oncology clinical trials presented in this chapter, it is also possible to shed light on the ethical concerns raised by information, comprehension, and voluntariness within oncology research's informed consent process,[3] a field of which RTD is part and with which it necessarily shares common ground.

Information, comprehension, and voluntariness were identified by the Belmont Report and have been adopted by every regulation on the topic as the three basic features required for informed consent in research to be valid and ethical. However, they reveal their problematic abstractness when translated into informed consent forms for oncology research. As a consequence, *formal* requirements can be applied in many different ways when translated into the research setting, and this openness may jeopardize patients' ability to make decisions that truly reflect their preferences. It thus becomes essential to further discuss these requirements in order to understand what sort of complications they raise within a concrete oncology research context.

When implemented in the concrete oncology setting, information, comprehension, and voluntariness—the basic features of an informed consent process identified by the Belmont Report—raise ethical issues relating to informed consent in both its *formal* and *substantial* senses. In the following sections, I present some ethical issues concerning informed consent for oncology research by focusing on information, comprehension, and voluntariness; in doing so, I consider how these issues apply to RTD.

5.5   Information: Ethical Issues

Informed consent is the most fundamental institution in the conduct of clinical trials. It is challenging to create a consent form that not only complies with the regulatory requirements of a correct *formal* sense, but also gives scientific information in a way that is comprehensible to a layperson (Pandiya 2010).

---

[3] It is worth noting that the problematic aspects of informed consent in research, which I will be raising, refer to adults and competent individuals who are able to choose what they consider to be best for themselves. My analysis will not consider individuals who have limited decision-making capacity (whether from birth or from a specific event in their life) and minors. These categories raise important issues; however, their particular condition requires an extensive and separate discussion that is not at the core of this book's aim.

In oncology research, besides giving medical information to the patient, the informed consent form[4] must also describe scientific details such as trial design, possible risks and benefits, treatment options, and the right to withdraw. According to the Belmont Report, the information that should be included in the informed consent form comprises the "research procedure, their purposes, risks and anticipated benefits, alternative procedures (where therapy is involved), and a statement offering the subject the opportunity to ask questions and to withdraw at any time from the research" (Belmont Report 1979). However, this general norm does not provide adequate details about the quality and the quantity of information an informed consent should provide, aspects that are of real concern within an informed consent form. With regard to quality, it is frequently difficult to establish the level of "science" or technical details that should be offered as part of informed consent. The result is that informed consent forms are generally replete with complicated scientific terms and procedures that the non-expert may find hard to comprehend. With regard to quantity, the length of informed consent forms is not easy to determine, since the form must be sufficiently short so that patients have the time to read it, yet it also needs to contain all relevant information. No limit is officially stated concerning the maximum number of pages of an informed consent form. As a result, informed consent forms tend to be particularly long in order not to omit any relevant information, yet they also frequently fail to provide the quality of information required by a patient to make an informed decision. The risk is that informed consent forms end up becoming a pile of information written in small characters to meet the legal responsibilities of medical staff. In other words, informed consent, if not adequately structured, fails to serve as an effective tool to help patients make their informed decisions. Instead, it becomes a document full of incomprehensible information aimed at avoiding legal actions, and one that no patient is likely to read thoroughly.

Language is frequently an issue in informed consent forms. Often, the terminology is too technical and scientific, making it difficult for the layperson to read and understand (Beskow and Weinfurt 2019). Moreover, the language may be poor and unclear. Sentence structure and grammar can contain mistakes, in particular when informed consent forms and information sheets are inaccurately translated from another language. According to Mader and Playe's (1997) study on informed consent readability, there is a positive correlation between potential risk to the subject and the length of the informed consent document. Mader and Playe highlighted that, as the potential risks of study participation increase, the reading difficulty of the consent form increases to the extent that high-risk forms require more than a "high school level of reading skills for comprehension" (Mader and Playe 1997).

How information is presented plays a key role in the decision-making process. The structure of the information—that is, the way in which it is organized into sentences, paragraphs, and sections—may influence how individuals perceive the

---

[4]Within the discussion on information, comprehension, and voluntariness, the term "informed consent" applies to both the information sheet and the informed consent form.

information.[5] Informed consent forms should, therefore, require stricter standards of clarity in written communication. Specific guidance concerning the quality and quantity of information should be issued so that patients can truly rely on these forms to make a decision over enrollment in clinical trials.

Although further analysis goes beyond the scope of this work, in the section dedicated to information within the informed consent it is worth mentioning that when it comes to information requirements, patients' awareness of their own diagnosis is usually taken for granted. Nevertheless, in oncology this is not a trivial assumption. Physicians may conceal cancer diagnosis from patients, with or without the backing of the patient's family.[6] This has significant effects on oncology research and on informed consent for enrollment.

5.6 Comprehension: Ethical Issues

In relation to patients' comprehension of the informed consent for oncology research, it is important to discuss in detail why this requirement is so troublesome and hard to assess in patients. The framework that results from this discussion will also be valid for RTD.

Comprehension is a basic requirement for informed consent in research. It requires that the patient has the opportunity to read, consider, and understand the information presented in order to be able to make an informed decision. Although simple in theory, this requirement can be difficult to translate into practice (Beskow et al. 2015; Beskow and Weinfurt 2019). In assessing whether individuals comprehend the informed consent form, it is not possible to rely on the assumption that they either entirely comprehend or not. In most cases, individuals may report a partial comprehension—that is, they indicate that they did not understand all sections of the form equally. For this reason, assessing patients' comprehension across the whole informed consent form is challenging. Comprehension is not clear cut. Rather,

---

[5]Most importantly, the way in which information is presented may alter individuals' comprehension and consequently impair their judgment in relation to the informed consent. Given the relevant link between the structure of information and comprehension/voluntariness, the ethical discussion concerning these issues will be extensively addressed in the comprehension and voluntariness sections below.

[6]Disclosing medical information directly to patients, especially for cancer-related issues, has been widely adopted as an essential requirement for autonomy in Western society. However, this practice is not as recognized worldwide as we would expect it to be. However dangerous any type of generalization is, it must be recognized that in some Eastern cultural contexts medical professionals might be more inclined to withhold information from patients concerning their diagnosis, as not all patients want to receive extensive information about their medical conditions; moreover, families traditionally tended to conceal a terminal prognosis from their relatives in order to protect them. Despite common perceptions, this attitude can also be found in Western countries. The withholding of information about diagnosis and prognosis represents a significant ethical issue in oncology. For further information on truth telling in oncology, see, among others, Surbone (1992).

comprehension of informed consent for research is a process that entails the comprehension of several related elements. Patients have to comprehend first *that* they have to decide whether to consent or to refuse, and second *what* they are consenting to (or refusing). This clarification is not trivial: frequently, patients fail to appreciate that they have first to evaluate whether to participate. Patients often regard the consent form as something they *have* to sign. They overlook that the informed consent form contains the information they need to evaluate *whether* to sign or not, that is, to evaluate whether to participate or not in the trial. The second aspect that patients need to comprehend is *what* they are consenting to.

In considering comprehension, it is necessary to address the importance of an adequate assessment of benefits associated with the research. Such benefits vary, and an adequate appreciation of their differences is crucial for the informed consent process. How patients assess the risks and benefits associated with research plays a key role in the decision-making process, and this dimension of adequate understanding of risks and benefits also raises ethical concerns within the RTD framework.

Evaluation of the risks and benefits by those participating in research is a crucial step in the informed consent process in oncology, and hence it is an essential component of the ethical conduct of research. Only by being provided with information about the risks and benefits of a research trial is it possible for participants to understand and properly evaluate those potential risks and benefits, and hence to decide whether to enroll. For this reason, a proper risk/benefit assessment is required from the patient to avoid informed consent being critically impaired. Risks and benefits should be discussed between physicians and patients when the research is presented. After this conversation, patients should read the informed consent form in which a specific section should be dedicated to the risks and benefits associated with the study, thereby ensuring that the prospective participants have at their disposal adequate information on which to make a sound and voluntary decision concerning enrollment.

Unfortunately, informed consent is not in practice always carried out in this ideal way. The extent to which participants effectively consider and assess information concerning a study's risks and benefits remains unclear. In theory, patients provide their informed consent when they voluntarily choose to enroll in research, and this very consent implies an understanding and an assessment of different aspects of the study, including its risks and benefits. However, informed consent forms are typically long and use technical, complicated language, making them hard for prospective participants to understand. Research studies on patients' ability to understand important aspects of clinical trials as presented in informed consent forms, as well as studies on informed consent in oncology, have revealed that patients consistently have higher expectations about the benefits than are warranted.

5.6.1 Appreciation of Benefits

In oncology research, the adequate understanding of benefits associated with the trial is a relevant issue. Research aims to produce generalizable knowledge, rather than to provide patients with personal care (Fried 1974). Furthermore, several aspects of clinical research methodology—such as placebos and randomization—require a sacrifice of personal care for the sake of preserving the integrity of research data, and it is vital for prospective participants to understand this concept. The possibility (if any) and type of benefit available for a patient enrolled in a research trial are long-debated issues in the field.

Nancy King (2000) has outlined three different meanings of "benefit" in research, whose confusion is commonplace and problematic. Her analysis is a potentially valuable aid for patients' comprehension, so it should be implemented within the structure of informed consent forms. According to King, a "direct" benefit is one that is immediately perceived by the participant because of the treatment; a "collateral" benefit comes from participation in a study and is associated with altruism; and an "aspirational" benefit is related to benefits generated by science for future patients:

> *Direct* benefit to subjects, which is properly defined as benefit arising from receiving the intervention being studied;
>
> *collateral* benefit to subjects (the National Bioethics Advisory Commission calls this "indirect" benefit [Keyserlingk et al. 1995, pp. 327–328]),which is benefit arising from being a subject, even if one does not receive the experimental intervention (for example, a free physical exam and testing, free medical care and other extras, or the personal gratification of altruism);
>
> *aspirational* benefit, or benefit to society and to future patients, which arises from the results of the study. (King 2000, p. 333)

In addition to distinguishing between the meanings of "benefit" in research, King (2000) also describes three dimensions of potential benefit that are crucial to its assessment as well as to the disclosure to patients in order for the informed consent process to be valid. The dimensions, which can overlap to some extent, are the nature of the potential benefit, the magnitude of the potential benefit, and the likelihood of the potential benefit. The nature of the potential benefit refers to the sort of benefit expected. Magnitude refers to the size and duration of the potential benefit. Likelihood, the third dimension, refers to the real chances of benefit foreseen by the study. King's analysis is useful when considering how the informed consent process should be structured in order for patients to comprehend the probability (if any) of receiving benefit. To facilitate the informed consent process, it is essential for investigators to engage prospective participants in conversations concerning the nature, magnitude, and likelihood of potential benefits. These three dimensions of benefit could provide prospective patients with a clearer set of information concerning what they can reasonably expect. Only with this information can prospective participants evaluate whether the risk–benefit ratio is reasonable from their perspective.

In the informed consent process, patients need to understand the real chances of different benefits occurring. According to King's distinctions, clinical research is mainly designed to create "aspirational benefit", since its aim is to produce general-izable knowledge for future patients. Yet, an adequate appreciation of the different benefits at stake is problematic both for oncology clinical trials and for RTD. In oncology clinical trials, patients' misconception of benefits available is a relevant concern, first reported by Appelbaum et al. (1982) who termed it "therapeutic miscon-ception". Therapeutic misconception reflects the mistaken belief that decisions about a research subject's treatment will be based on that individual's condition and needs.[7] Patients affected by therapeutic misconception misconceive the nature of benefits associated with the research and maintain the expectation that, for instance, the research will provide them with direct rather than aspirational or collateral bene-fits, despite the research effectively being designed to create the latter rather than the former. This attitude is not rare in oncology research[8] and, by jeopardizing the informed consent process, it affects how patients understand both the procedure and the aim of clinical trials. In circumstances in which, for example, the patient assumes that the experimental drug might be the only chance of their survival, this cognitive cherry-picking makes prospective subjects believe in unrealistic outcomes by selectively accepting or ignoring information included in the consent form.[9]

A misconception of benefits may apply to RTD as well. As a form of postmortem research, RTD can bring no direct benefit to the prospective donor. Instead, the bene-fits associated with RTD are overtly aspirational, since they arise from the opportunity to help future generations of patients through the research performed on retrieved tissues. This aspiration might constitute a benefit for RTD donors, because they may derive comfort and solace from their altruistic action of donation. Nevertheless, misconception of benefits may arise with prospective donors (and their families). Patients might mistakenly perceive that they will be potentially reserved benefits upon enrollment in the procedure. For example, a prospective donor might think that enrollment will be "rewarded" with better end-of-life care or with some sort of privilege for them or their surviving family (see Chapter 8 below). This misconcep-tion may be fostered by the pressure that patients may feel to please, rather than to disappoint, their physician and the medical staff who may have research interests in cancer tissues.

Such misconceptions are extremely problematic and potentially impact on the informed consent process for RTD, so they need to be thoroughly considered. Since RTD involves postmortem research, the misconceived potential for collateral bene-fits—such as better end-of-life care upon enrollment—does not follow from the

---

[7] For further information on therapeutic misconception in clinical trials, see Appelbaum and Lidz (2008); Epstein and Lasagna (1969); Horng and Grady (2003); Joffe et al. (2001); Joffe and Weeks (2002); Park and Covi (1965); Penman et al. (1984).

[8] An interesting line has been traced between therapeutic misconception, misestimation, and opti-mism in oncology research. For further information on this difference, see Horng and Grady (2003); Lidz et al. (2015); Pentz et al. (2012).

[9] According to Appelbaum et al. (2004), therapeutic misconception can appear in two dimensions: as an incorrect belief that the needs of patients/participants will determine the kind of treatment assigned; and as an unreasonable appraisal of the likelihood of medical benefit from participation in the trial, caused by a misunderstanding of the nature of the research.

research itself—the donor will no longer be alive—but rather from the decision to consent to it. Thus, any benefits to the donor will arise before the research itself takes place, and not after. This contrasts with therapeutic misconception in other forms of oncology research, where misconceived direct benefits are associated with research results—and not with the patient's act of consenting to it—and are expected to occur subsequently. Thus, misconception of benefits is a concern in relation to all oncology research and its informed consent process, but it raises particular issues for RTD given that the misconception of benefits is most likely to arise in connection not with the research procedure itself but with the procedure for informed consent to that research.

5.6.2 Risk/Benefit Evaluation

Although an evaluation of risks is at the core of oncology clinical trials, such an evaluation applies only tangentially to RTD as a postmortem procedure. For this reason, risks, as well as benefits, will be addressed in this section in general, and by taking a perspective that regards patients' adequate assessment of these aspects as an ethical issue for both RTD and oncology research. Although the evaluation of risks and benefits is a fundamental requirement for patients to be able to make an informed decision, few studies have focused on how patients concretely make this assessment. Specific focus should be directed to understanding how prospective patients approach risk/benefit evaluation and the role (if any) that this evaluation plays in their decision to enroll or not.

Often, in the clinical setting, the only knowledge patients have about a study's risks and benefits is the one they have gathered via physicians or investigators during the informed consent process. In such cases, the information received by patients has been filtered by medical staff and has not been directly assessed by patients. However, patients with cancer seldom assess risks and benefits before enrolling in a clinical trial for oncology research. Ulrich et al. (2016, p. 7) investigated this aspect of oncology trials by conducting a study of 110 cancer clinical trial participants (Phase I, II, III) at a large cancer center in the United States. They found that most of the interviewed patients (66%) reported that they "thought the benefit of participating made taking the risks worth it, but half of the cohort said they didn't assess the risks and benefits".

Patients, in particular those, such as many cancer patients, who are severely ill, are connected to their physicians by a deep feeling of trust. From the patient's perspective, physicians are supposed to be more than reliable sources of medical information: a physician's recommendation or advice is seldom called into question by patients, since it is widely recognized that physicians want the best for their patients. This assumption is, however, problematic. Leaving aside the potential issue of physicians' conflicting interests—for example, a physician might obtain research benefits from a patient enrolling in a specific clinical trial or from a patient's donation of tissues in the case of RTD—a physician's reporting of risks and benefits to a patient will inevitably reflect a set of personal values and analysis that may impair patient evaluations. Physicians and patients might consider the trial from different perspectives. The physician might perceive the trial through a "scientific" lens, according to which the ultimate aim is to produce generalizable knowledge. Patients, on the other hand, are likely to perceive the trial through a personal, patient-centered lens. In the case of RTD, physicians and patients may evaluate tissue donation differently according to their religious, cultural, social, and personal beliefs, leading to a different weighting of risks and benefits. This may extensively impact the resulting evaluation of enrollment.

Even if risks are not central to RTD as a postmortem procedure, there are relevant considerations in the evaluation process that might have a psychological impact on patients and families. The decision to enroll in RTD or not can lead to extreme stress and anxiety about the retrieval, and this might be exacerbated by conflicting religious, cultural, and personal views. Given that the decision to enroll in RTD involves an extremely personal dimension, it is even more important that is assessed by patients during the decision-making process.

When information concerning risk/benefit assessment is obtained only through the filter of a physician, then the informed consent process might be impaired. It is, of course, vitally important that a patient relies on a trusted physician for medical advice. However, when assessing risks and benefits in order to evaluate enrollment, direct risk/benefit information should be obtained in order to gain an unfiltered assessment, thereby ensuring a thorough informed consent based on the values and preferences of a patient rather than those of the treating physician.

Cancer patients are generally reported to feel severely vulnerable, and this vulnerability might explain a tendency to entrust themselves to others, to delegate decisions, and to rely on the advice of physicians or third parties in order to make evaluations and final decisions. It is widely recognized that cancer patients might be affected by "decision fatigue" (Levy 2012), a condition that leaves them feeling tired because of aggressive therapies, surgeries, and general weakness. This condition can affect their willingness to engage in a decision-making process; for this reason, these patients are likely to seek assistance when evaluating their options. Thus, cancer patients may make the choice in the company of their physician, before they hear an explanation of the risks and benefits, and then simply decide to stick to their chosen course of action regardless of additional information they are provided with. According to Jansen (2014), people who have already made a decision they consider important are in an "implementation mind set" and are, therefore, less likely to think about the risks and benefits associated with that particular choice. Thus, patients who have already decided to take the advice of their physician to enroll in research will be less interested in a thorough risk/benefit evaluation since they have already made their choice and are focused on following their course of action. This problem may also apply to RTD.

It is clear, therefore, that an assessment of risks and benefits, and of the nature of benefits, by patients in cancer research and in RTD is replete with ethical dilemmas. The relevance that personal, religious, and cultural values acquire within the RTD framework make a proper assessment particularly important. It is vital that a weighting of the information associated with RTD be evaluated by the prospective donor according to their personal framework of values, rather than by a physician whose own values and preferences may filter (even if subconsciously) the information and impact the informed consent process. The importance of a personal evaluation by the patient is central and should be encouraged by a specifically trained ethicist whose role is to engage with the patient, assist them throughout the informed consent process, and facilitate the emergence of the patient's preferences.

5.7 Voluntariness: Ethical Issues

Codes and regulations require informed consent to be voluntary. Nevertheless, assessing the voluntariness of consent and tracing its boundaries is difficult in practice. The informed consent process is rife with potentially influencing factors. The oncology setting is an environment in which interference becomes ethically concerning in relation to patients' informed consent. Since voluntariness is a broad concept, it is necessary to narrow the focus in order to define what is meant by this term. I will use the same meaning of voluntariness that Beauchamp and Childress (2013, p. 138) adopt, according to which a person acts voluntarily if "he or she wills the action without being under the control of another person or condition".

The relationship between patients and physicians is fraught with ethical controversies and poses critical concerns in relation to patients' voluntariness. The power imbalance between patients and physicians may create obligations and invisible bindings that are hard to dissolve and whose repercussions may affect the decision-making process. In RTD, for instance, patients may feel the pressure to consent to tissue donation in order to please their physician, and they may perceive that their level of end-of-life care will be jeopardized by a refusal to enroll in an RTD program. In oncology clinical trials as well, patients' feeling of weakness caused by the disease, along with the perception of restricted treatment options, may play a significant role in medical decisions, as Berg et al. (2001, p. 68) have pointed out: "[A] patient suffering from a life-threatening disease may feel as though she has little choice regarding treatment. Physicians should be aware of how vulnerable patients may be to the coercive influence of unrealistic hope, especially those suffering from chronic, life-threatening disorders."

The implications of this assessment require further analysis. Diseased patients, in particular cancer patients, may feel vulnerable given the impact of the disease on their lives. They need to interact with new physicians, in unfamiliar settings, and their future is uncertain. In these circumstances of likely discomfort, stress, and anxiety, patients can be subject to influences that take different forms and have different intensities.[10]

It is commonly recognized that research involves several potential conflicts of interest. The most obvious issue in clinical ethics is that clinical research presents an intrinsic conflict of interest when the clinician is also a researcher. A clinician-researcher has an obligation not only to particular patients but also to perform accurate research according to a protocol. These two duties potentially conflict. Physicians have a significant influence on patients due to their knowledge and expertise. However, when physicians are at the same time investigators, their potential influence on patients might increase dramatically. A patient's underlying trust that the physician/investigator will act for their good—a trust that is often based on a positive interpersonal relationship with the physician—may hamper the patient's ability to pursue their personal interests effectively. When patients are advised by their physician to enroll in an oncology trial that the physician is leading, the patient might feel obligated to follow that advice in order not to disappoint the physician. Similarly, patients in RTD programs may be reluctant to refuse to donate tissues when this chance is offered by their treating physician who is also a researcher and may have scientific interests in the patient's tissues. The patient might consent based on the belief that doing so will gain their physician's respect and will ensure that they continue to receive the same level of care and attention.

In addition, families might significantly influence patients' decisions on research enrollment. Overwhelming circumstances of pain and stress caused by the disease may induce patients to please their family by acceding to their wishes, even if this results in a partial deprivation of autonomy.

Difficulties in the decision-making process are also related to the patient's information processing. Typically, patients are overexposed to a huge amount of information, often at the same time, concerning their illness and the available treatment options, and this information is frequently couched in unfamiliar terms. The problem of information overexposure is exacerbated by time pressures impinging on many informed consent discussions, particularly those involving different physicians using scientific and technical terms that may generate disorientation, anxiety, and discomfort among patients.

These factors play an important role within decision-making. It is essential for patients to make informed decisions that reflect what they truly consider best for themselves. The concept of voluntariness is broad and full of implications, so it is essential to narrow the focus on the aspects that are relevant to the present discussion. Beauchamp and Childress (2013) highlight three forms of influence: coercion,

---

[10]However, this is not to say that cancer patients, because of their medical condition, are inherently unable to decide what is best for themselves. Cancer patients can voluntarily choose to let someone decide on their behalf, should they realize that they themselves do not want to choose from among the medical choices.

manipulation, and persuasion.[11] Of these, manipulation has the greatest implications for informed consent.

5.7.1 Manipulation

Manipulation is a broad category that includes any "intentional and successful influence of a person by non coercively altering the actual choices available to the person or by non persuasively altering the person's perceptions of those choices" (Faden and Beauchamp 1986, p. 261). In the medical setting, the most dangerous form of manipulation is informational manipulation, which involves deliberately presenting information in a way that alters an individual's understanding of a specific circumstance in order to induce them to act according to the manipulator's will.

In the informed consent process, informational manipulation is a significant concern, since it may impair the decision-making process: withholding or concealing information and lying may lead a patient to believe in false assumptions that affect their decision. Investigators can put more or less emphasis on specific aspects of the research, and risk/benefit information is an especially delicate area. For this reason, it can be problematic when the assessment is made only through the physician, without any direct assessment by the patient, as has been argued previously in this chapter. How medical staff present prospective patients with information about research—the manner in which they frame it through choice of sentence structure, tone of voice, and gestures—can affect patients' perceptions. For example, a patient's reaction is likely to be different if information is presented in a positive way—for example, "this therapy succeeds most of the time"—than if it is presented negatively—for example, "this therapy fails in 30% of the cases" (Weinfurt et al. 2003). Such phrasing choices play a key role in oncology clinical trials, in which the presentation of the probability of a direct benefit—and, thus, the patient's perception of this probability—may significantly impact an enrollment decision.

Likewise, the framing and perception of probability of benefits are also important in the patient's RTD decision-making process. Within the framework of RTD, expression of probability overtly regards aspirational benefits, namely, the chances that donated tissue will make a concrete contribution to the treatment of future cancer patients. When a cancer patient evaluates the option to donate tissues after death for research aims, the belief that there are high chances of benefit for scientific research

---

[11] Coercion occurs when one party "intentionally and successfully influences another by presenting a credible threat of unwanted and avoidable harm so severe that the person is unable to resist acting to avoid it. The three critical features in this definition, at least for our purposes, are that: 1. The agent of influence must intend to influence the other person by presenting a severe threat; 2. there must be a credible threat, and 3. the threat must be irresistible". Persuasion is the "intentional and successful attempt to induce a person, through appeals to reason, to freely accept - as his or her own - the beliefs, attitudes, values, intentions, or actions advocated by the persuader" (Faden and Beauchamp 1986, p. 261). Coercion and persuasion will be not further addressed in this book.

may increase the likelihood of consent. Further complicating perceptions of probability is that there are, according to Hacking (2002), "belief-type" and "frequency-type" statements about probability. Belief-type statements are an expression of the level of confidence a person feels about a statement—for example, "I am 20% sure I will need a car by the end of the year". Frequency-type statements are statements of fact, usually based on data concerning the "relative frequency of some episode occurring in the long run in the population"—for example, "50% of people need a car". When the real chance of success associated with clinical oncology trials, especially those in the early phases, is discussed, assumptions are based on outcomes observed over many trials using a frequency-type probability. Frequency-type expressions are relevant for a patient's comprehension within the informed consent process. However, many patients are likely to think of probability in a belief-type way, and this may result, in the informed consent process both for oncology research and for RTD, in a different perception of the probability of benefit. As Weinfurt et al. (2003) point out:

> Imagine that a patient is explicitly told that 5% of a population of patients is likely to experience a benefit in a phase I trial. Now the patient is asked: "On average, how many patients out of 100 will benefit from a phase I trial?" This utterance (question) is phrased explicitly in frequency-type terms, and thus any answer other that "5" would be considered an error of memory or comprehension. Such a response could justifiably be taken as reason for an ethical concern that the patient has not been informed to the point of understanding.

> But consider that the same patient, having been informed by means of frequency-type probability statements, is asked, "What is the chance that you will benefit from a phase I trial?" Because a frequency-type expression cannot apply to a single episode, the patient must make a response that takes a form other than a frequency-type statement. The only other form available is a belief-type statement in which the patient states his or her confidence that a single episode – experiencing success – will occur. The mismatch between the type of uncertainty the investigator is using (frequency) and the type of uncertainty the patient is using (belief) makes it impossible to judge the correctness of the patient's response. After all, even if a patient understands that 5 out of every 100 patients will benefit from the treatment, the patient may still be 90% confident that he or she will be 1 of these 5.

How information is framed remains extremely important for determining patients' expectations of a research trial and their decision to enroll in it or not. It becomes necessary in the healthcare setting to monitor situations in which misleadingly presented information—whether as a result of manipulation or due to trivial inaccuracy—may prevent patients from making judgments that reflect what they believe to be best for themselves.

5.8 Conclusion

Ethical quandaries raised by informed consent for oncology research relate to information, comprehension, and voluntariness, which provide a significant framework for understanding and addressing the ethical issues raised by RTD. By using this framework, this chapter has presented issues that are relevant to a consideration of the informed consent process for RTD.

First, the importance of information, especially in relation to the quality and quantity of the information presented in the consent form, is a recurring problem that has crucial implications for patients' understanding and appreciation of what is at stake, and hence for their decision-making. Furthermore, the voluntary nature of consent is also relevant in RTD. Enrollment should be free from influences stemming from family dynamics or from an unbalanced patient–physician relationship.

Concerning comprehension, a central aspect of both general oncology research and RTD is an estimation of the benefits (including the types of benefits) and risks associated with a program, because a misinterpretation of the benefits at stake could impair the informed consent process. The adequate appreciation of the nature of the benefit at stake is a crucial step in the decision-making process, and this aspect should not be passively transferred from the physician to the patient. Rather, this assessment constitutes the basis for the evaluation of enrollment, which should take into account personal, cultural, and religious values. Ethical concerns associated with the informed consent are profound and dense and should not be underestimated.

Given this, well-structured informed consent forms and information sheets designed to meet the patient's needs are essential for ensuring an adequate informed consent process. In addition, a better structured oral conversation with the physician to discuss the intervention at stake is also necessary. However, all these measures should be accompanied and implemented by a clinical ethicist, a key figure who can provide patients with discreet and competent support to ensure that the patient's decision-making in the healthcare setting reflects the patient's preferences and choices. Only through the assistance, support, and mediation of the clinical ethicist, whose functions will be discussed in the next chapter, can the ethical concerns associated with information, understanding, and voluntariness of the informed consent process be adequately addressed.

References

Appelbaum, P.S., et al. 2004. Therapeutic misconception in clinical research: Frequency and risk factors. *IRB* 26 (2): 1–8.

Appelbaum, P.S., and C.W. Lidz. 2008. The therapeutic misconception. In *The Oxford handbook of clinical research ethics*, ed. E.J. Emanuel et al., 633–644. Oxford: Oxford University Press.

Appelbaum, P.S., et al. 1982. The therapeutic misconception: Informed consent in psychiatric research. *International Journal of Law and Psychiatry* 5: 319–329.

Beauchamp, T.L., and J.F. Childress. 2013. *Principles of biomedical ethics*, 7th ed. Oxford: Oxford University Press.

Berg, J., et al. 2001. *Informed consent: Legal theory and clinical practice*. Oxford: Oxford University Press.

Beskow, L.M., et al. 2015. Informed consent for biobanking: Consensus-based guidelines for adequate comprehension. *Genetics in Medicine* 17 (3): 226–233.

Beskow, L.M., and K.P. Weinfurt. 2019. Exploring understanding of "understanding": The paradigm case of biobank consent comprehension. *American Journal of Bioethics* 19 (5): 6–18.

Churchill, L.R. 2005. Toward a more robust autonomy: Revising the Belmont Report. In *Belmont revisited: Ethical principles for research with human subjects*, ed. J.F. Childress et al., 111–125. Washington: Georgetown University Press.

Declaration of Helsinki. 1964. https://www.wma.net/policies-post/wma-declaration-of-helsinki-ethical-principles-for-medical-research-involving-human-subjects/.

Emanuel, J.E., et al. 2008. *The Oxford textbook of research ethics.* Oxford: Oxford University Press.

Epstein, L.C., and L. Lasagna. 1969. Obtaining informed consent: Form or substance. *Archives of Internal Medicine* 123: 682–688.

Faden, R., and T.L. Beauchamp. 1986. *A history and theory of informed consent.* Oxford: Oxford University Press.

Fried, C. 1974. *Medical experimentation: Personal integrity and social policy.* Amsterdam: North Holland Publishing.

Grodin, M.A. 1992. Historical origins of the Nuremberg code. In *The Nazi doctors and the Nuremberg Code: Human rights in human experimentation*, ed. G.J. Annas and M.A. Grodin, 121–144. Oxford: Oxford University Press.

Hacking, I. 2002. *An introduction to probability and inductive logic.* Cambridge: Cambridge University Press.

Horng, S., and C. Grady. 2003. Misunderstanding in clinical research: Distinguishing therapeutic misconception, therapeutic misestimation, and therapeutic optimism. *Ethics and Human Research* 25: 11–16.

Howard-Jones, N. 1982. Human experimentation in historical and ethical perspectives. *Social Science and Medicine* 16 (15): 1429–1448.

Jansen, L.A. 2014. Mindsets, informed consent, and research. *Hastings Center Report* 44 (1): 25–32.

Joffe, S., et al. 2001. Quality of informed consent in cancer clinical trials: A cross-sectional survey. *Lancet* 358: 1772–1777.

Joffe, S., and J.C. Weeks. 2002. Views of American oncologists about the purposes of clinical trials. *Journal of the National Cancer Institute* 94: 1847–1853.

Jones, J.H. 1993. *Bad blood: The Tuskegee Syphilis experiment.* Mumbai: The Free Press.

Keyserlingk, E.W., et al. 1995. Proposed guidelines for the participation of persons with dementia as research subjects. *Perspectives in Biology and Medicine* 38 (2): 319–361.

King, N.M.P. 2000. Defining and describing benefit appropriately in clinical trials. *The Journal of Law, Medicine & Ethics* 28 (4): 332–343.

Levy, N. 2012. Forced to be free? Increasing patient autonomy by constraining it. *Journal of Medical Ethics* 40 (5): 293–300.

Lidz, C.W., et al. 2015. Why is therapeutic misconception so prevalent? *Cambridge Q Healthcare Ethics* 24 (2): 231–241.

Mader, T.J., and S.J. Playe. 1997. Emergency medicine research consent form readability assessment. *Annals of Emergency Medicine* 29 (4): 534–539.

Moreno, J.D. 1999. *Undue risk: Secret state experiments on humans.* New York: W.H. Freeman.

Nuremberg Code. 1947. https://www.ushmm.org/information/exhibitions/online-exhibitions/special-focus/doctors-trial/nuremberg-code#Permissible.

Pandiya, A. 2010. Readability and comprehensibility of informed consent forms for clinical trials. *Perspectives in Clinical Research* 1 (3): 98–100.

Park, L.C., and U. Covi. 1965. Nonblind placebo trial: An exploration of neurotic patients' responses to placebo when its inert content is disclosed. *Archives of General Psychiatry* 12: 336–345.

Penman, D.T., J.C. Holland, et al. 1984. Informed consent for investigational chemotherapy: Patients' and physicians' perceptions. *Journal of Clinical Oncology* 2: 849–855.

Pentz, R.D., et al. 2012. Therapeutic misconception, misestimation, and optimism in participants enrolled in phase I trials. *Cancer* 118: 4571–4578.

Surbone, A. 1992. Truth telling to the patient. *JAMA* 268: 1661–1662.

Ulrich, C.M. 2016. Cancer clinical trial participants' assessment of risk and benefit. *AJOB Empirical Bioethics* 7 (1): 8–16.

US Department of Health and Human Services, Office for Human Research Protection. n.d. Code of Federal Regulations, 45 CFR 46.102(d). https://www.hhs.gov/ohrp/regulations-and-policy/regulations/45-cfr-46/index.html#46.102.

US Department of Health and Human Services, US Food and Drug Administration. 2018. Clinical research versus medical treatment. https://www.fda.gov/forpatients/clinicaltrials/clinicalvsmedical/default.htm.

US Department of Health and Human Services, Office for Human Research Protection. 1979. The Belmont Report: Ethical principles and guidelines for the protection of human subjects of research. https://www.hhs.gov/ohrp/regulations-and-policy/belmont-report/index.html#xinform.

Weinfurt, K.P., et al. 2003. Patient expectations of benefit from phase I clinical trials: Linguistic considerations in diagnosing a therapeutic misconception. *Theoretical Medicine and Bioethics* 24 (4): 329–344.

Chapter 6
The Clinical Ethicist

Abstract Advances in biotechnology and medicine invariably have ethical implications. New approaches to healthcare and treatment options are now possible, but they are accompanied by unprecedented ethical and clinical questions that require thorough analysis. These issues involve extremely delicate areas at the intersection of patients, patients' families, and healthcare professionals. To address clinical issues and their associated ethical quandaries, there is a need for an ethical framework that can help guide the decision-making process, resolve conflicting interests with ethical implications, and balance the principles of autonomy, beneficence/non-maleficence, and justice within the clinical setting. Ethical considerations have long played a crucial role in the healthcare setting. Clinical ethics is especially relevant in the field of oncology, no more so than in relation to Rapid Tissue Donation (RTD). The dedicated and adequately trained RTD ethicist plays a key role in a patient's decision-making process, especially in ensuring that informed consent is understood in both its *formal* and *substantial* meanings.

6.1 The Clinical Setting and the Need for Guidance

Advances in scientific knowledge and technology have been accompanied by crucial changes in how healthcare decisions are made and managed. These changes have impacted all parties involved: patients, patients' families, and healthcare professionals. The new possibilities in healthcare and treatment options raise unprecedented ethical and clinical questions. These questions, which arise on a daily basis, are often embedded in clinical encounters between patients and healthcare professionals and take place within settings that are unknown and unfamiliar for most patients. It is crucial, therefore, that ethical considerations relating to the new possibilities are thoroughly analyzed.

The new possibilities opened up by progress in the medical field has not only expanded the quantity of available solutions but also their quality. Evaluating the healthcare options requires focusing not only on how to treat and when, but also on whether to treat. When do the risks associated with an intervention outweigh its benefits? How should the best interest of the patient be evaluated? At the same time, the legal code, norms, and guidelines are increasingly intersecting with the governance of issues relating to care choices and end-of-life decisions. Clinical issues with relevant ethical implications may be related to, among other areas, withholding or withdrawing therapies, medical futility, informed consent, therapeutic misconception, an adequate risk–benefit evaluation, and the concept of the patient's best interests. These issues invariably involve highly delicate topics, debates, and decisions that pertain to the perspectives of patients, families, and healthcare professionals. To address the ethical quandaries raised by clinical issues, there is a need for an ethical framework that can guide the decision-making process, mediate conflicts, and balance the principles of autonomy, beneficence, and justice within the clinical setting. Clinical ethics, in the various forms it has taken over time, aims to guide the approaches and decisions of all involved in healthcare practice, to mediate between parties, and to resolve ethical dilemmas through careful analysis and mitigation.

There is no universally accepted definition of clinical ethics. Leading figures in the field agree on the importance of clinical ethics, but they have offered various definitions of it.[1] Nevertheless, these different definitions do reveal some common traits. According to Boyd's (1997, p. 40) contribution to *The New Dictionary of Medical Ethics*, clinical ethics is "a form of applied ethics practiced in the hospital or healthcare setting and concerned with actual clinical choices. It may involve a clinical (or hospital) ethics committee whose functions include ethics policy making, education and case consultation, and/or a clinical ethicist who works alongside staff." Hurst (2012), in the *Encyclopedia of Applied Ethics*, maintained that the role of the clinical ethics committee is to consider the ethical dimensions of patient care. Fletcher (1991, p. 859) defined it as an "interdisciplinary activity to identify, analyze and resolve ethical problems that arise in the care of particular patients. The major thrust of clinical ethics is to work for outcomes that best serve the interests and welfare of patients and their families". A different perspective was taken by Jonsen et al. (1992, p. 1) who defined clinical ethics as a "practical discipline that provides a structured approach to decision making that can assist physicians to identify, analyze, and resolve ethical issues in clinical medicine". A further definition by LaPuma (1990, p. 321) described clinical ethics as a

> process of identifying, analyzing and resolving moral problems of a particular patient's care. The primary goal of clinical ethics is to improve patient care with bedside assistance. Clinical ethics seeks to improve the relationship between the patient and the clinician and the relationships among patients, family, and hospital.

The importance of a clinical ethics perspective within healthcare was also emphasized by the President's Commission for the Study of Ethical Problems in Medicine and Biomedical and Behavioral Research (President's Commission) established by the US Congress in 1978. In a report published in 1983, the Commission recognized the need for a clinical ethics dimension that encompasses three main tasks: (1) an ethical analysis of clinical cases of particular concern; (2) the creation and

---

[1] There are many definitions of clinical ethics; here, I will discuss only some of them.

use of guidelines and recommendations to address ethical problems; and (3) promotion and management of education and training to promote ethical awareness among healthcare professionals (President's Commission 1983).

Despite the lack of a univocal and universally accepted definition, clinical ethics is usually understood today to refer to an emerging field that focuses on the decision-making process in the healthcare setting, the protection of the rights and preferences of patients and other parties involved in healthcare, and the encouragement of cooperation between stakeholders. According to the American Medical Association (AMA 1998), clinical ethics consultations should aim to support "informed, deliberative decision making on the part of patients, families, physicians, and the health care team", and clinical ethicists perform the role of "helping to clarify ethical issues and values, facilitating discussion, and providing expertise and educational resources [and they] promote respect for the values, needs, and interests of all participants, especially when there is disagreement or uncertainty about treatment decisions" (AMA 1998, 10.7.1). The same AMA code outlines a detailed set of aims and conduct that should guide a clinical ethics consultant,[2] according to which the consultant should:

(a) Seek to balance the concerns of all stakeholders, focusing on protecting the patient's needs and values.

(b) Serve as advisors and educators rather than decision makers. Patients, physicians, and other members of the care team, health care administrators, and other stakeholders should not be required to accept the consultant's recommendations. Physicians and other institutional stakeholders should explain their reasoning when they choose not to follow the consultant's recommendations in an individual case.

(c) Inform the patients when an ethics consultation has been requested (if the request was not made by the patient or family) and seek patients' agreement to participate. Ethics consultants should respect the decision of a patient or family not to participate, whether that decision is indicated formally through explicit refusal or informally by not taking part in discussions.

(d) Respect the rights and privacy of all participants and ensure that appropriate steps are taken to protect the confidentiality of information disclosed in the consultation.

(e) Have appropriate expertise or training—for example, familiarity with the relevant professional literature, training in clinical/philosophical ethics, or competence in conflict resolution—and relevant experience to fulfill their role effectively.

(f) Adopt and adhere to policies and procedures governing ethics consultation activities in keeping with medical staff bylaws, including accountability and standards for documenting the consultation in the patient's medical record.

---

[2]The AMA lists requirements for physicians who provide consultations. However, the requirements considered here are relied on for trained clinical ethicists who may not have a background in medicine.

(g) Ensure that all stakeholders have timely access to consultation services in nonemergent situations and as feasible for urgent consultations (AMA 1998, 10.7.1).

Above all, the profound changes that have revolutionized medicine in recent decades, and which have stemmed not only from scientific and technological progress but also from cultural, social, and economic developments, have created the need for clinical ethics as a guide to analyze ethical dilemmas, concerns, and conflicts that arise in the healthcare setting. The need for clinical ethics is all the more apparent given the growing awareness that modern medicine challenges long-established paradigms by posing quandaries that medical science cannot solve on its own.

6.2   The Birth of Ethics Committees to Address Clinical Issues with Ethical Implications

The aims and structure of clinical ethics consultations have evolved significantly in recent decades. From the 1960s to the 1980s, scientific and technological developments generated unanticipated clinical questions that were rife with ethical implications. Organ transplants, allocation of scarce resources, terminal illness, and end-of-life care, among others, required adequate solutions and policies. It became clear that the appropriateness and quality of these processes would benefit from heterogeneous perspectives and expertise. To this end, not only clinicians, but also professionals from other backgrounds, such as philosophy, law, biomedicine, and theology, became increasingly involved in the analysis of emerging ethical issues.

One of the earliest and most famous examples of this heterogeneous approach to addressing clinical issues with ethical implications arose in 1960 when a physician at the Seattle Artificial Kidney Center invented the shunt to make chronic dialysis feasible (Jonsen 1998). This invention changed the lives of many patients affected by kidney failure. However, the treatment was extremely expensive, and the nine-bed capacity of the center could not meet the demands of the numerous patients applying for treatment. Thus, an ethical question was raised: How should such a scarce life-saving resource be allocated? In order to address this dilemma, the King County Medical Society constituted two different committees to evaluate and select applications. The first committee, which was composed of physicians, was intended to shortlist candidates by focusing on their clinical and psychiatric profile. Shortlisted candidates then went through a further selection process by the second committee in which there was an accurate case-by-case analysis of each candidate's psychological, economic, and social aspects, such as age, background, education, "past performance and future potential" (Jonsen 1998, p. 212). This second committee, charged with the extremely delicate task of deciding how to allocate the life-saving resources— and, in effect, of deciding who would live and who would die—consisted of members from a variety of backgrounds, such as a lawyer, a businessman, and a homemaker.

The diverse composition of this committee was an important recognition of the value that different expertise and perspectives can bring to the solution and analysis of a problem that, although emerging within the boundaries of the clinical context, has important repercussions along other dimensions. The multidisciplinary structure of the committee and the mixture of different professional profiles dealing with the complex ethical implications of clinical problems is the horizon toward which the development of clinical ethics committees has subsequently been oriented.

These earliest committees, which were established in the twentieth century, were intended to help in the management of healthcare decisions that had ethical implications. Typically, the ethics committees—such as those for dialysis—consisted of small groups, not always interdisciplinary in composition, that aimed to address ethical concerns relating to specific topics. In the US in the 1920s, sterilization committees were formed to evaluate which individuals with mental impairments should undergo forced sterilization; likewise, abortion selection committees functioned before the 1973 Supreme Court decision that legalized abortion (Post et al. 2007, p. 13). In the 1960s, the US Institutional Review Boards were established to monitor and watch over research with human subjects after the discovery of abuses during medical experiments.

As scientific and technological advances accelerated, the range of ethical issues relating to high-tech care also increased, with the result that hospitals began constituting ethics committees to manage a wider array of such issues. These committees were established to provide stakeholders with guidance on the management of clinical issues with ethical implications arising within the healthcare setting. Over time, such committees have "taken on the additional functions of staff education, clinical guideline development, institutional policy advisement, and case review. Some ethics committees also advise on resource allocation and express or reinforce the institution's commitment to certain values" (Post et al. 2007, p. 14).

Discussion of clinical cases that pose ethical problems is a widespread practice within the healthcare context. The nature of such discussions and the regulatory framework in which they occur vary significantly from one country to the next. But with technological development and the widening of care, what remains constant is the need to resolve the increasing number of clinical issues with ethical implications.[3]

6.3 Core Ethical Principles

Applied ethics in the clinical setting, regardless of its structure, assumes an undeniable relevance in all situations in which the deepest, most personal and intimate values of the person come into play, such as in cases of suffering, disease, and end

---

[3] Among the further literature on this, see: Ahronheim et al. (2000), Annas (1991), Arras et al. (1999), Fletcher (1991), Lo (2000), Mappes and Degrazia (2001), Miller (1995), Moreno (1998), Rosner (1985), Ross et al. (1986, 1993); Rothman (1991); Thomasma (1993); Thomasma and Monagle (1998).

of life. Clinical ethics are central to the understanding and management of critical moments in the life of the patient and to analyzing and settling the moral dilemmas that arise when competing values come into conflict.

Any discussion of applied bioethics involves an overview of its core ethical principles. Such foundations guide and underpin clinical ethics consultations at all levels and in whatever form they take. The ethical principles by which any clinical ethicist must abide include respect for persons, which consists of encouraging patients to exercise their choices through a decision-making process that enables their wishes and preferences to be respected; beneficence through the promotion of the patient's best interest; non-maleficence in order not to cause harm; and justice in the form of a fair allocation of benefits and a focus on the healthcare dimension.

The principle of respect for persons, according to the Belmont Report (1979), finds its concrete application in two core requirements: the requirement to "acknowledge autonomy and the requirement to protect those with diminished autonomy". The principle of respect for persons thus promotes a sound decision-making process that reflects patients' preferences and choices in their healthcare, while simultaneously safeguarding the best interests of those whose decision-making capacity is limited. According to the Belmont Report, the concrete application of the principle of respect for persons is informed consent, which is characterized by participation, shared decision-making, and respect for the autonomy of the patient. Only when the individual cannot make decisions are others asked to make decisions on the patient's behalf. Respect for persons requires that "subjects, to the degree that they are capable, be given the opportunity to choose what shall or shall not happen to them. This opportunity is provided when adequate standards for informed consent are satisfied" (Belmont Report 1979, Part C Paragraph 1).

In order to foster respect for persons in their healthcare choices through informed consent, three aspects should be kept in mind: information, comprehension, and voluntariness. A more detailed discussion of these aspects has been proposed in Chapter Five above. In the discussion that follows in this chapter, only general aspects that are essential to deepening the role of the clinical ethicist will be considered.

In order to provide their consent, subjects should be given specific and complete information about the medical intervention they are about to undergo. Part of this information might include details regarding the schedule of the entire medical procedure, the anticipated risks and benefits, available alternatives to the procedure, and costs. However, as discussed in Chapter 5, even the most comprehensive and detailed list does not give a complete picture of this requirement, because a fundamental aspect of information regards how this knowledge is received and comprehended by the subject. For this reason, information is often a problematic requirement as it is not easy to establish a standard for evaluating how much and what kind of information should be given to the subject.

Comprehension is inextricably tied to information. In order for the consent process to be accurate, patients must comprehend the information they are given. However, the level of comprehension that subjects achieve is not easy to assess. Informed consent is not a simple signature on a document or a matter of a patient's intelligence and rationality. Comprehension is also affected by the context: poor exposition, limited time available to discuss details, and complex language may often be reasons for inadequate comprehension. Special assistance is required when a patient has severely limited comprehension, such as in cases of immaturity or mental impairment.

There is also a requirement that the consent be voluntary and free from external pressures, coercion, or undue influence. Coercion occurs when an "overt threat of harm is intentionally presented by one person to another in order to obtain compliance", and undue influence occurs through an offer of "excessive, unwarranted, inappropriate or improper reward or other overture in order to obtain compliance" (Belmont Report 1979, Part C Paragraph 1). The complexity of assessing subjects' voluntariness is connected to the nuanced nature of external pressures. It is hard (and, perhaps, even impossible) to precisely establish where justifiable persuasion—such as when it is applied ostensibly to benefit the subject—ends and undue influence begins. It is clear that when assessing the voluntariness of minors or persons with diminished autonomy, the level of complexity increases significantly.

Patients' decisions do not take place in a closed world; rather, contextual features, such as family, culture, religion, and patient–physician relationships, have significant repercussions on medical choices. Thus, patients' choices might be blurred by a wide range of external factors. The influence of external factors potentially threatens to prevent patients from developing choices that reflect their personal preferences, inclinations, and values.

In scenarios where there are significant ethical concerns and possible conflict between ethical interests, clinical ethics consultants can foster a sound decision-making process to protect patients from harmful external pressures. Ethics consultations involve professional figures whose objective is to analyze ethical concerns and conflict in order to facilitate a process in which shared solutions are shaped so that they respect the interests and values of patients, and are the best fit for all parties involved in the case. Ethical issues are embedded in every clinical encounter between clinicians, patients, and their families. Such encounters seldom involve decisions about what is wrong and what is right; rather, they are a matter of finding the most reasonable solution among various possible options. Moreover, clinical ethics analysis rarely focuses only on a single individual; instead, medical decisions involve not only individual patients but also their families, loved ones, as well as medical staff. The analysis therefore has to consider the personalities, values, histories, and beliefs of several parties. Thus, clinical ethicists aim to facilitate shared solutions and to mediate among individuals, interested parties, and conflicting interests by using a discreet but firm process. Although it is usually appropriate to honor and safeguard the wishes of a patient, it is also important to consider the ethical principles that give rise to other, often concurrent, obligations.

The principles of beneficence and non-maleficence have profound implications for healthcare professionals and caregivers, whose long-established mission is to heal and care. According to the principle of beneficence, persons are to be "treated in an ethical manner not only by respecting their decisions and protecting them from harm, but also by making efforts to secure their well-being" (Belmont Report 1979, Part B Paragraph 2). The principles of beneficence and non-maleficence imply an adequate evaluation of the benefits and risks involved in healthcare interventions.

Well-being, benefits, risks, and harm for a specific patient are not purely scientific values; they are also values informed by preferences, expectations, standards, and perspectives, and hence they can vary widely. Consequently, standards of beneficence and non-maleficence might differ among physicians, patients, and healthcare professionals. Recognizing such differences is extremely important in the healthcare setting, because many conflicts can arise out of these differences. In light of this, mediation by the clinical ethicist is crucial for pinpointing the origins of conflicts of interest and for guiding relevant parties toward a resolution.

The principle of justice is of great relevance for medical ethics. This is especially the case with distributive justice, according to which benefits and burdens should be fairly allocated across the population. In the ethical meaning of this principle, ethically well-founded reasons should govern decisions relating to why certain individuals have access to specific benefits while others do not, such as in the allocation of scarce resources. The role of the clinical ethicist is vital to ensuring the adequate management of conflicts triggered by such complicated issues in healthcare.

The different ways in which people evaluate decisions about healthcare are strongly impacted by their beliefs, inclinations, fears, attitudes, and values, which are in turn shaped by their culture, religion, education, and social background. These differences can lead to communication barriers, power imbalances, misunderstandings, and conflicting perspectives that potentially challenge relationships between patients, healthcare providers, and other involved parties by negatively impacting on clinical choices. As the clinical setting becomes increasingly filled with choices that have dramatic ethical implications, the role of clinical ethics consultations becomes ever more important.

6.4 Different Kinds of Clinical Ethics Consultations

Healthcare ethics consultations respond to a local need to help patients, families, and healthcare professionals address ethical concerns arising in clinical settings. Ethics consultations are widely considered to be a relevant part of care, and they are offered as a permanent service within healthcare structures in numerous countries. The consultations provide a rigorous method of case-based analysis, and they create a space for discussion in which relevant issues and emerging conflicts are unraveled and resolved. Such a process respects, on the one hand, the values and preferences of patients and families so that they feel adequately cared for, and, on the other hand, it enables healthcare providers to resolve conflicts by relying on a set of shared principles and standards.

Ethics consultations are not always offered through ethics committees. Rather, ethics consultations adapt to and are shaped by the requirements of healthcare settings, and they can be structured in different ways according to particular needs and, of course, depending on available resources. Traditionally, the three most common ways to perform an ethics consultation are by an individual ethics consultant, a whole ethics committee, or an ethics consultation team, each of which has advantages and disadvantages.

In the individual ethics consultant model, one person is assigned to perform the consultation. Since the consultant is working alone, scheduling a meeting is not particularly complicated, and the entire organizational and administrative process is relatively straightforward. By relying on a particularly agile structure, this model manages to guarantee prompt and rapid assistance and is best suited to intervene in urgent cases. Furthermore, this model favors the development of a strong interpersonal relationship between the consultant and the parties involved, as the same consultant can follow the development of a specific case and can be regarded as a reference point for stakeholders. However, while the agile structure of a single consultant is one of the model's strengths, it can also present disadvantages. In particular, the consultation and the associated analysis relies on only one point of view, with limited interaction with peers, so it does not benefit from a multidisciplinary perspective. Hence, in this model there is an increased risk of the consultant's own values and biases intruding into the analysis of the case. These negative aspects might, however, be mitigated by the consultant being encouraged to confer with colleagues to discuss any limits in their analysis. Indeed, even the most experienced and qualified consultant always benefits from a discussion with third parties or colleagues for the resolution of the most complex cases.

The second model of clinical ethics consultation is the ethics committee. In this model, an interdisciplinary committee, typically numbering 6–20 members, performs the ethics consultation. A major advantage of this approach is its interdisciplinarity and the multiplicity of perspectives from which the case analysis can benefit. Given the heterogeneity of the professional profiles that generally constitute the committee, third parties will be offered the opportunity to confer with a wide range of professionals by requesting a single consultation. However, the articulated structure of the ethics committee means that it can be complicated to organize and schedule a meeting. Consequently, this solution is rarely suitable for urgent requests for consultations or for those that require short notice. It is also worth considering that a large number of white-coated professionals might not achieve the desired effect of putting involved parties at their ease, and that it may be counterproductive: patients, families, and third parties may feel intimidated, with the result that mediation and any necessary conflict resolution might be difficult to achieve.

In the ethics consultation team model, a small group of qualified consultants chosen for their competencies conducts the meeting. This model represents an intermediate solution between the two models discussed above. The advantages of the team model involve the ability to ensure that an analysis is built on diverse perspectives. Despite its more agile structure, the disadvantages associated with this model partially reflect those of ethics committees and are mainly associated with the complex organization of meetings involving multiple professionals, especially when short notice is required for urgent consultations.

Given that all three models have strengths and weaknesses, there is no best model for all cases. The choice of model should be based on a case-by-case evaluation in order to determine the most appropriate form of consultation. Regardless of the model chosen, healthcare ethics consultations should, as a general rule, be easily accessible and available for patients, families, and healthcare providers, and they should not be limited only to acute care hospitals. Where an ethics consultation is offered, in whatever form it takes, clear indications should be given concerning what the service offers, how to access it, and what to expect.

6.5   Approaches to Clinical Ethics Consultations

Conducting an ethics consultation is never straightforward since it involves extremely delicate and sensitive matters. Should consultation be required to resolve a conflict between the parties involved or to assist in a decision-making process, the work of consultants requires high levels of sensitivity as well as solid training and expertise. Above all, consultations cannot be improvised or happen randomly; they must be requested by an involved party, whether the patient, a family member, or a member of the healthcare staff. Upon request, a consultation should be conducted according to a rigorous method that enables the analysis and the whole process to flow harmoniously and to involve the active participation of all concerned parties.

The methods used in the consultations are manifold. As a general rule, no single method is better than another, as long as the chosen method is thorough and enables a case analysis based on mediation and respect toward the parties involved. Nevertheless, because consultation methods can differ considerably, some methods are more suitable than others for resolving certain issues. The chosen method also depends on the consultation model. It is up to the expertise and flexibility of consultants to determine the most appropriate method or to adapt the method to the circumstances of the case.

An example of a method commonly relied on is the CASES approach. This is a systematic method for guiding ethics consultations by using standard steps that can be adapted for different situations. CASES is an acronym that stands for each step of the process, which are explained further below:

Clarify the consultation request
Assemble the relevant information
Synthesize the information
Explain the synthesis
Support the consultation process (Fox et al. 2005).

Clarify the Consultation Request
This first step of the process functions as a filter and is aimed at understanding and elucidating the circumstances in which the request for consultation emerged. Understanding what exactly happened before consultants enter the scene is crucial, not least because it serves to determine whether the case is suitable for an ethics

consultation and whether or not it will benefit from the process. If not, the case should be referred to the competent service or office as soon as possible.

After determining that the requested consultation actually involves an ethical question and that it can therefore be analyzed using the CASES approach, precise information concerning the circumstance in which the consultation emerged should be gathered. This should lead to the central focus of this first step: to formulate the ethics question to address. A well-set question is key to a sound analysis.

Assemble the Relevant Information
In this step, the consultant aims to gather information to answer the ethical question underpinning the case. Four categories of information should be gathered: first, salient medical facts, such as the patient's medical records, advance directives, and other relevant documents; second, the patient's preferences and interests, which should be sought, if possible, by interacting directly with the patient in order to understand the their circumstances; third, the preferences and interests of other relevant parties in order to gain further insights and put a more detailed picture together; and fourth, ethics knowledge consisting of an overview of codes and regulations, as well as of similar cases that might help in the analysis.

Synthesize the Information
Once the information has been gathered and put together, the consultant synthesizes it by using analytical skills and moral reasoning in order to answer the ethics question. This process involves choosing the most suitable approach based on the circumstances of the case and on the needs of the parties involved.

Explain the Synthesis
Once the synthesis has been reached, it should be expressed to the parties involved through direct and clear communication in order for them to be actively involved in the process. The development of the process should be noted on health records as it unfolds.

Support the Consultation Process
The final step is to support and manage the ethics consultation process in order that ethical concerns are addressed properly. It is crucial that the involved parties are on the same page.

The CASES approach is a thorough step-by-step model that guides consultants through active cases. It is a highly linear method and constitutes an effective and easy-to-remember scheme for conducting consultations that are simultaneously rigorous and fluid.

Another method that is frequently used is bioethics mediation (Dubler and Liebman 2004). Adopted by the Montefiore Medical center, this was one of the very first ethics consultation services to have been developed. The bioethics mediation model for consultation relies on techniques for governing and resolving conflicts by facilitating discussion among involved parties. The method attempts to identify and delineate conflicts; to level the playing field so that imbalances and disparities of power between the parties involved are limited; to help parties become aware so that they can identify, and successfully communicate, their interests; to encourage the search for common ground for discussion; and to foster, guide, and follow up discussions in order to resolve any conflict (Post et al. 2007, p. 151). Bioethics mediation and bioethics consultation are slightly different approaches:

> Bioethics consultation refers to a directed substantive process. The consultant listens to the parties and helps move them toward a principled resolution of the dispute by explaining ethical principles and legal rules, applying them to the facts, and presenting the social consensus on the permissibility of different practices. Bioethics mediation refers to the use of classical mediation techniques to identify, understand, and resolve conflicts. Bioethics mediation and bioethics consultation may both be employed in a particular case at different points in the process. Mediation is more inclusive and empowering, and consultation is more authoritarian and hierarchical; either or both may be required in any complex case, even within a single meeting. (Dubler and Liebman 2004, p. 14)

Besides the specific circumstances and complexities of each case, the main goal of the bioethics mediator is to guide parties to a resolution of a conflict that is agreed by all, while also ensuring that solutions fully respect the preferences and values of those involved. These preferences and values might include religious beliefs, cultural attitudes, and personal differences. Unlike the CASES approach, a notable advantage of this method is its intrinsic flexibility. Indeed, bioethics mediation can be applied to a wider range of cases that present ethical conflicts. Despite its flexibility, however, bioethical mediation should be recognized as not being the most suitable method for all circumstances. Indeed, there are some cases where its implementation is not recommended, such as when "the conflict is out of control before it comes to the attention of the mediator"; when a "psychological problem or psychiatric diagnosis affects one of the parties and is at the heart of the disagreement [such that] reason and argument will be ineffective because of illness and distortion"; and when "outsiders may have an interest in augmenting conflict [where the use] of legal process and the press to make political points dooms private reconsideration and resolution" (Post et al. 2007, p. 152). Apart from these specific cases, however, the approach and method exemplified by bioethics mediation has the advantage of using reasoning and a dialectic of confrontation to guide the parties involved toward a shared solution that respects the values of all stakeholders.

There are many other methods and processes that can regulate or inform clinical consultations. It is important that consultants, relying on their expertise and training, adopt a rigorous method that suits the specific case and the particular interests of those involved.

Regardless of the model used, the delicate role performed by clinical ethicists in consultations in the healthcare setting requires appropriate training, qualifications, and expertise. Today, robust and thorough training programs are available for clinical ethicists, but this has been a relatively recent development. In 1998, the American Society for Bioethics and Humanities (ASBH), in response to its identification of a lack of adequate education and training programs, published a *Code of Ethics and Professional Responsibilities for Healthcare Ethics Consultants*. The code sets out the core ethical responsibilities of individuals performing healthcare ethics consultations. It addresses the ethical concerns of persons involved in health-care decision-making and delivery. The ASBH listed the following requirements for ethics consultants:

1. Be Competent.
2. Preserve Integrity. Healthcare Ethics Consultants should consistently act with integrity in the performance of their role.
3. Manage Conflicts of Interest and Obligation. Healthcare Ethics Consultants should anticipate and identify conflict of interest and obligation, and manage them appropriately.
4. Respect Privacy and Maintain Confidentiality. Healthcare Ethics Consultants should protect private information obtained during consultations, handling such information in accordance with standards of ethics, law, and organizational policy.
5. Contribute to the Field. Healthcare Ethics Consultants should participate in the advancement of Healthcare ethics consultations.
6. Communicate Responsibly. When communicating in the public arena (including social media), Healthcare Ethics Consultants should clarify whether they are acting in their role and should communicate in a manner consistent with the norms and obligations of the profession.
7. Promote just Health Care within Healthcare Ethics Consultations. Consultants should work with other healthcare professionals to reduce disparities, discrimination, and inequities when providing consultations (ASBH 1998).

The requirements listed by the ASBH constitute an especially relevant foundation for general guidelines for healthcare ethics consultants. However, the need and importance of coordinating recognized training programs does not follow immediately. According to a survey carried out by Fox et al. (2007), although ethics consultations were widespread in the US, guidelines regarding ethics-specific training were lacking. Only 5% of ethics consultation providers had completed a fellowship or graduate degree program in ethics or bioethics, whereas 41% had learned to perform ethics consultations with formal, direct supervision by an experienced member of an ethics consultation service, and 45% had learned independently without formal supervision. The 2007 survey reveals that individuals working in ethics consultation services in hospitals with no academic affiliation were considerably more likely to have gained experience and skills without official supervision. In academically affiliated hospitals, this was less frequent. Similarly, individuals involved in ethics consultations in private or for-profit hospitals were more likely to have learned without formal supervision than were those working in non-profit or religiously affiliated hospitals (Fox et al. 2007). Moreover, a significant proportion of those who performed ethics consultations in US hospitals were part of the clinical staff, most commonly physicians (34%) and nurses (31%). Furthermore, 11% of consultants were social workers and 10% were chaplains, but consultants were rarely non-clinicians (4%), such as attorneys, philosophers, or theologians (ibid.).

Fox et al.'s (2007) survey indicates that on-the-job experience and practice are important. However, although experience is vital in this area, it cannot replace academic training. The survey's findings revealed that ethics consultants had a concerning lack of knowledge in bioethics, an area of knowledge that, in order to perform their role and meet patients' needs, they should have been familiar with. Ethics consultants are an important component of the healthcare system, so there is a significant need for clear standards relating to accountability and educational training for these professionals. Moreover, it is important to develop criteria to evaluate whether these standards are met.

It is important to emphasize that the clinical ethicist should not be perceived as a replacement or a substitute for the physician or the healthcare staff, and that there is no overlap between their respective roles. Rather, each figure should harmoniously cooperate within the healthcare setting by mutually respecting their specific roles. Clarifying the importance of cooperation and the mutually supporting (rather than overlapping) roles is fundamental, as the perception of an overlap could be counter-productive and lead to further conflicts, when the role of clinical ethics is instead to resolve conflicts.

6.6   The Clinical Ethicist in the Oncology Setting

Within the oncology setting, be it research or care, ethical dilemmas arise on a daily basis. These dilemmas may relate to, among other things, the goals of care, therapeutic misconception, decision-making, appreciation of risks and benefit, the informed consent process, and the transition to palliative care. Such issues, and notably the transition to palliative care, are emotionally laden and require careful consideration of the patient's best interests. Although these challenges are not exclusive to the oncology setting, they nevertheless acquire a specific relevance when associated with cancer care.

There are many reasons for this special bond, most of which are at the intersection between the personal and clinical dimensions. The relationship of care between clinical staff and a patient battling malignancy is particularly intense and frequently involves families and relatives. The oncology setting is rife with ethical issues that concern circumstances in which there has to be management of moral dilemmas and conflicting values between the involved parties. Cancer represents the plague of our times, and although the general perception of this disease has shifted somewhat thanks to more effective treatments, the mortality long associated with this disease has meant that it has long been a taboo. Clinical ethics consultations emerged within this context and have evolved to assist in mediating between parties through careful guidance and in mitigating ethical issues that arise in relation to decision-making about end-of-life issues, research trials, and treatment options.

Cancer care involves challenging decisions, interpersonal conflicts between involved parties, communication barriers, ethical dilemmas, and psychological distress and anxiety that require careful and sensitive management. Ethics consultants play a unique role in assisting the involved parties, using a reasoned approach to respect patients and their families and facilitating a relationship between patients and their healthcare providers. In oncology, there is no preferred form of ethics consultation. It might be performed by a group or by a single consultant. Whatever form the consultation takes, it is essential that an ethics consultant is available to patients, their families, and healthcare professionals.

6.7 The Role of the RTD Ethicist

Given the fundamental role of clinical ethics within the healthcare setting, and specifically in relation to oncology, it is unsurprisingly just as crucial in relation to Rapid Tissue Donation (RTD). Clinical ethics is particularly relevant for the RTD decision-making process and for supporting patients, families, and healthcare providers in their evaluations. Moreover, it has a vital role in mediating between parties should conflicts about choices and values arise. Although RTD shares some ethical issues with wider oncology research, the particular framework of RTD involves some unique concerns that have to be addressed. The standard training of a clinical ethics consultant may be inadequate to meeting the needs of the parties involved. For this reason, it is recommended that specific training be provided for ethics consultants who support RTD programs. This training is designed to equip ethics consultants with an understanding of the specific RTD framework and with the skills required to provide valid support for RTD stakeholders.

It is especially important that the consultant has RTD ethical expertise for several reasons. First, it is vital for patients to interface with the same person given the sensitivity of the issues and the multi-stage structure of the informed consent process for RTD, as will be discussed in the next chapter. Patients and families should be able to engage in dialogue with the same consultant who can act as a point of reference for them. They should be able to consult with the RTD ethicist if doubts arise, or if they need further information or clarifications about the procedure. In other words, they should be able to establish a relationship with one person on whom they can rely throughout the process. Moreover, given the urgency and short notice by which patients might be requested to express interest in RTD within the healthcare setting, it is important to rely on agile consultation structures that can intervene promptly, rather than on teams composed of several members who might be unable to schedule the consultation swiftly.

The ethicist in the RTD context plays a key role in the informed consent process. This role is not that of generic clinical ethics support in the field of oncology care or oncology research. RTD ethicists should receive adequate training and be dedicated and committed to RTD programs. While they still perform their standard functions of support in the decision-making process, in mediation, and in conflict resolution, these functions take place in the specific circumstances created by RTD. For this reason, and precisely because of the delicate and particular issues that RTD can raise for patients and families, the RTD ethicist will take over at a particular moment to support the patient's decision-making process and to ensure, as far as possible, that the patient's choices reflect their personal attitudes, expectations, and values. The RTD ethicist is trained in promoting communication between patients and their families concerning the decision to enroll in the program, as well as in encouraging communication on delicate topics such as genetic data and the ends to which retrieved samples might be used. It is important that the RTD ethicist has the skills to tailor the conversation about RTD to the specific target audience, and that they should be extremely sensitive to cultural, religious, and personal values. A thorough mastery of language (sometimes the choice of words makes a vital difference in a conversation) is required in order to modulate the conversation according to the needs and attitudes of patients and families. Moreover, the RTD ethicist is the connecting element between patients, families, and medical staff should any hurdle or conflict arise. The role of the RTD ethicist is to mediate between parties in order to promote a healthy environment and to mitigate the distress and the anxiety that end-of-life decisions bring up. Given that the RTD ethicist responds to specific requirements (Camporesi and Cavaliere 2019) and should not rely on improvisation, their specific training should include preparing them with the professionalism to adequately assist and support families in the extremely delicate moment after the patient's death when the retrieval process takes place.

While offering research opportunities in cancer care, RTD involves a remarkable ethical component for prospective patients and their families, as well as for medical staff. Within this framework, the ethics consultant specifically trained in RTD aims to provide ethical support relating to a plethora of issues, dilemmas, and uncertainties that may arise with regard to the procedure. The value of the RTD ethicist within an RTD program reflects the importance and relevance of ethics within the medical field. The ethicist should inform and guide relevant parties, and they need to be equipped with the training and skills to discuss the complex ethical issues that arise in the medical setting in a respectful and non-judgmental way. In particular, they should be able to place the patient and their values at the core of the decision-making process.

In relation to the main topic of this work—informed consent for RTD in both its *formal* and *substantial* senses—the comprehensive aim of the RTD ethicist is to promote a thorough informed consent process within the framework of the (mostly) unprecedented ethical issues RTD brings up. Within this framework—which will be discussed in further detail in the next chapter—the RTD ethicist is the key figure for enabling informed consent, according to both its meanings, to become a reality for RTD.

References

Ahronheim, J., J.C. Moreno, and C. Zuckerman. 2000. *Ethics in clinical practice*, 2nd ed. New York: Aspen Publishers.

AMA (American Medical Association). 1998. Code of medical ethics opinion. https://www.ama-assn.org/delivering-care/ethics/ethics-consultations.

American Society for Bioethics and Humanities, Task Force on Standards for Bioethics and Humanities. 1998. Core competencies for health care ethics consultation: The report of the American Society of Bioethics and Humanities.

Annas, G.J. 1991. Ethics committees: from ethical comfort to ethical cover. *Hastings Center Report* 21 (3): 18–21.

Arras, J.D., B. Steinbock, and A.J. London. 1999. Moral reasoning in the medical context. In *Ethical Issues in Modern Medicine*, 5th ed, ed. J.D. Arras and B. Steinbock, 1–40. California: Mayfield.

Belmont Report. 1979. https://www.hhs.gov/ohrp/regulations-and-policy/belmont-report/read-the-belmont-report/index.html.

Boyd, K.M. 1997. Clinical ethics. In *The new dictionary of medical ethics*, ed. K.M. Boyd, R. Higgs, and A.J. Pinching, 40. London: BMJ Publishing Group.

Camporesi, S., and G. Cavaliere. 2019. We cannot all be ethicists. *Nature* 575 (7784): 596.

Dubler, N.N., and C. B. Liebman. 2004. *Bioethics mediation: A guide to shaping shared solutions.* United Hospital Fund.

Fletcher, J. 1991. The bioethics movement and hospital ethics committees. *Maryland Law Review* 50: 859–894.

Fox, E. et al. 2005. *Ethics consultation: Responding to ethics concerns in health care.* Veterans Health Administration. www.va.gov/integratedethics/download/EthicsConsultationPrimer.pdf.

Fox, E., S. Myers, and R.A. Pearlman. 2007. Ethics consultation in United States hospitals: A national survey. *American Journal of Bioethics* 7 (2): 13–25.

Hurst, S. 2012. Clinical ethics. In *Encyclopedia of applied ethics*, 2nd ed, ed. R. Chadwick, 476–487. Amsterdam: Elsevier.

Jonsen, A.R. 1998. *The birth of bioethics.* Oxford: Oxford University Press.

Jonsen, A.R., et al. 1992. *Clinical ethics*, 3rd ed. New York: McGraw-Hill.

LaPuma, J. 1990. Clinical ethics, mission and vision: Practical wisdom in health care. *Hosp Health Serv Adm* 35: 321–326.

Lo, B. 2000. *Resolving ethical dilemmas: A guide for clinicians*, 2nd ed. Philadelphia: Lippincott Williams & Wilkins.

Mappes, T.A., and D. Degrazia. 2001. *Biomedical ethics*, 5th ed. New York: McGraw Hill.

Miller, B. 1995. Autonomy and the refusal of life-sustaining treatment. In *Ethical issues in modern medicine*, 4th ed, ed. J.D. Arras and B. Steinbock, 202–211. California: Mayfield Publishing.

Moreno, J.D. 1998. Ethics committees and ethics consultants. In *A companion to bioethics*, ed. H. Kuhse and P. Singer, 475–484. Oxford: Blackwell.

Post, L.F., J. Blustein, and N.N. Dubler. 2007. *Handbook for Healthcare Ethics Committees.* Baltimore: John Hopkins University Press.

President's Commission for the Study of Ethical Problems in Medicine and Biomedical and Behavioral Research. 1983. Final Report on Studies on the Ethical and Legal Problems in Medicine and Biomedical and Behavioral Research.

Rosner, F. 1985. Hospital medical ethics committees: A review of their development. *JAMA* 253 (18): 2693–2697.

Ross, J.W., et al. 1986. *Handbook for Hospital Ethics Committees.* Chicago: American Hospital Publishing.

Ross, J.W., et al. 1993. *Health Care Ethics Committees: The next generation.* Chicago: American Hospital Publishing.

Rothman, D.J. 1991. *Strangers at the bedside: A history of how law and bioethics transformed medical decision making.* New York: Basic Books.

Thomasma, D.C. 1993. Assessing bioethics today. *Cambridge Quarterly of Healthcare Ethics* 2: 519–527.

Thomasma, D.C., and J.F. Monagle. 1998. Hospital ethics committees: Roles, membership, structure, and difficulties. In *Health care ethics: Critical issues for the 21st century*, ed. J.F. Monagle and D.C. Thomasma, 460–470. New York: Aspen Publishers.

Chapter 7
Informed Consent for RTD: An Overview

Abstract This chapter sets out an informed consent process, in both the *formal* and the *substantial* senses of that term, for the innovative technique of Rapid Tissue Donation (RTD). The analysis focuses on practical aspects that have relevant clinical ethics implications, such as choosing the right timing for the first discussion and selecting the right spokesperson. The proposed solution relies on the physician who manages the treatment and on the RTD ethicist whose objective is to ease the decision-making process and to mediate any conflicts that arise within the RTD setting. The structure of the informed consent process consists of different phases in order to encourage, as far as possible, a choice that reflects the patient's values and preferences. Although the informed consent process is intended for adults with decision-making capacity, some consideration will also be given in this chapter to minors and to adults with limited decision-making capacity. Aspects pertaining to informed consent for data processing are also discussed; in addition, implications for blood relatives with regard to the informed consent process for RTD are considered.

7.1 Introduction

Drawing on the evidence of oncology clinical trials and through a consideration of the ethical issues raised by Rapid Tissue Donation (RTD), the objective of this chapter is to develop an informed consent for RTD in two senses of the term: first, an informed consent in a *formal* sense, whereby a legally recognized dimension is constituted for cancer patients who wish to donate tissues; and second, an informed consent in a *substantial* sense, whereby the requirements are modeled according to the autonomous authorization of the cancer patient.

RTD is known as "warm" tissue donation because of the short window between time of death and collection of tissue samples during the autopsy. This window is fixed at between two and six hours after death to allow high-quality tissues to keep their molecular landscape optimal for research aims. Given this, RTD involves a very innovative and particular kind of research, whose features have significant ethical, clinical, and logistical implications.[1] Many implications raised by RTD recall those of general oncology research; however, due to its innovative character, RTD involves new, unexplored aspects whose impact on patients and their relatives remains challenging. One of the main limitations is that the narrow window for RTD retrieval imposes a rigid schedule that does not leave time for discussion and decision-making. Thus, logistical limitations imposed by RTD raise ethical concerns as families will have less time to spend with their loved ones immediately after death and bodies will need additional time to be returned to families for burial ceremonies because of tissue collection. These features of RTD may present significant difficulties to families, particularly if they disrupt burial plans or religious traditions, and so they could cause additional distress for grieving families.

In light of these considerations, communication between healthcare professionals and patients and between healthcare professionals and families is especially important within informed consent for RTD.[2] Organ donation, transplantation literature, and experience provide a potentially useful framework for understanding how patients and families can be approached for discussion about donation.[3] However, it may be dangerous to rely excessively on the tradition of organ transplantation as a guide. Organ transplantation and RTD may be similar in certain respects, and the practice of organ donation may, for example, offer useful indications and guidelines for understanding how patients and families perceive discussions and make decisions about postmortem donation. But in other respects, organ donation and RTD are extremely different in their aims and scope. As a consequence, overlap between the two procedures should be subject to a thorough evaluation.

Studies in bioethics have generally found that communication about donations entails significant emotional distress for families as well as for clinical staff (Galushko et al. 2012; Horne et al. 2012). Such distress is caused by the sensitive nature of the subject of death, which usually generates strong emotional reactions that make it difficult to discuss; moreover, death has different nuances depending on the religious and cultural background of patients, families, and clinical staff.[4] Thus, the discussion with patients and families about decisions relating to what will happen after death requires training, sensitivity, and an understanding of how ethnic, cultural, and

---

[1] See also: Alsop et al. (2016); Embuscado et al. (2005); Hulette et al. (1997); Laposata (2017); Lindell et al. (2006); McIntyre et al. (2013); Rubin et al. (2000); Schabath et al. (2013); Shah et al. (2004); Spunt et al. (2012); Van der Linden et al. (2014).

[2] See also: Achkar et al. (2016); Bryant et al. (2015); D'Alessandro et al. (2008); DuBois and Anderson (2006); Fernandez et al. (2012); Galushko et al. (2012); Hanto et al. (2005); Horne et al. (2012); Hyde and White (2009); Krueger and Casey (2015); McVearry-Kelso et al. (2007); Merchant et al. (2008); Pentz et al. (2003); Pentz et al. (2005); Quinn et al. (2013); Rodrigue et al. (2006); Sque et al. (2005); Stevens (1998); Stouder et al. (2009); Thombs et al. (2005); Wilkinson (2005); Williams et al. (2003); Willis and Draper (2012); Wood et al. (2004).

[3] See also: the Genotype-Tissue Expression (GTEx) Project, ELSI sub-study run by Smirnoff et al. (http://www.siminoffresearchgroup.org/research-studies/gtex/) and GTEx Consortium (2013) The Genotype-Tissue (GTEx) project, *Nature Genetics* 45(6):580–585.

[4] The wish for bodily integrity of the deceased, in order to accommodate an open casket funeral for example, can be very meaningful for some patients and families. Such issues will be more fully discussed in Chapter 8.

religious attitudes might affect decision-making. Although the literature on organ donation potentially offers significant help in this regard, the aims of organ donation for transplantation are significantly different from those of RTD, and it is important to be aware of this in discussions with patients. Whereas organ donation aims at transplantation leading to a real, immediate, and concrete translation into a new life, RTD has research purposes that bring far less immediate and concrete help for other individuals in terms of new treatment development. Nevertheless, the literature on and collected experience of organ donation for transplantation represents a valid and solid starting point for comprehending some of the issues posed by RTD. These ethical implications will be at the core of this chapter, which aims to propose a path for an informed consent process to govern RTD. To meet this aim, this chapter presents a step-by-step analysis of the main issues encountered along such a path within the oncology setting.

7.2 Step I—Picking the Right Timing for the First Discussion

Timing is of great concern when it comes to approaching patients to discuss RTD. Oncology patients refer to physicians because they are fighting their disease; this means they are looking for and expecting treatment, hope, assistance, and survival. This expectation makes it hard for healthcare providers to initiate a discussion with patients about what to do with their bodies after death. Presenting patients with the option of postmortem donation of their tissues may result in a dangerous misunderstanding and a loss of hope in patients whose aim is to fight and survive cancer. It is important that all patients, no matter the stage of their disease, feel that they are cared for as patients whose lives are at stake, and not as bodies with precious biological samples to be retrieved for research (Quinn et al. 2013). For this reason, approaching the discussion is subject to possible misinterpretations and entails dealing with extremely delicate and sensitive issues that can evoke emotional reactions. Patients may interpret the RTD discussion as an indirect way adopted by healthcare providers to tell them they are not going to survive the disease, which is precisely what must be avoided in this situation. A patient's needs must be at the center of care, and RTD should be a parallel matter to be evaluated only with the patient's agreement. Choosing the right timing for the RTD discussion amounts, therefore, to a delicate decision that must be given careful consideration.

7.2.1 At the Moment of Diagnosis

Since the RTD program we are focusing on is reserved for cancer patients only, the first option could be to indifferently present every cancer patient at the moment of diagnosis with the chance to donate tissues after death, no matter the stage of their disease or their chances of survival. This path would avoid discriminating between those patients at an advanced stage whose survival is more critical and those at earlier stages. Presenting the RTD option at the outset of care may contribute to a perception that this discussion is a standard step within the entire treatment procedure, and not one that is necessarily associated with the severity of the disease. This choice would help patients become familiar with the subject without associating it with the degeneration of the disease, since the discussion would be conducted in the same way with patients in the early stages of the disease as it would with terminal patients.

Despite its concrete advantages, a discussion of RTD at the point of diagnosis and without discrimination can present several challenges. The moment of diagnosis is overwhelming for patients and families who have just learned the news but have not assimilated it yet. Presenting them with the option to donate tissues after death may not be the most appropriate choice in such circumstances, since patients are understandably often confused and scared when receiving their diagnosis. At a moment when life has suddenly changed for a patient, it seems inappropriate to introduce further delicate issues for them to think about. Thus, it could be premature to talk about RTD at this time, especially because patients expect healthcare staff to follow a diagnosis by providing them with hope and a therapeutic plan, rather than with options that relate to what will take place after their death. Patients expect their treatment and survival to be at the core of healthcare providers' attention, so death should not even be mentioned at this point. In the immediate period after cancer diagnosis, patients and families are in the initial and delicate stage of establishing and developing a vital relationship of trust with the physician. Any discussion of RTD at this point may be misinterpreted as a diversion of attention away from the patient's primary needs for the sake of research and may thus significantly impair and jeopardize the relationship between physician and patient. As McIntyre et al. (2013 p. 6) reported, a cancer diagnosis is a delicate moment for patients and families in which they depend on healthcare providers' care and attention, and hence patients and their families may be distressed by anything that seems to divert from this aim: "Why talk about donating tissue if they have just been diagnosed with cancer? Excuse me, I don't want to talk about donating organs or tissues because we're going to fight this."

7.2.2 At the Moment When Cancer Becomes Terminal

In light of the above, it seems that the first meeting with the oncologist is not the best time for an RTD discussion. A second option is to initiate the RTD discussion when a patient's illness is terminal—that is, when they have exhausted available treatment options and the focus has moved largely to palliative care. In these circumstances, patients may be presented with the chance to help future patients by donating cancer tissues to research. Patients might regard RTD as a way of helping others when they are no longer able to help themselves.

Nevertheless, this option also presents challenges. The point at which cancer patients realize they no longer have survival chances might be a delicate one, because they have to cope with their poor prognosis. In such an overwhelming moment, in which patients' perception of the future has irremediably changed, it could come across as tasteless and insensitive for healthcare providers to propose the option of tissue donation after death. It is worth noting that entering the terminal stage of a disease often does not constitute a precise moment. Instead, it is a path that, little by little, can unfold in that direction, but it is not a sudden change that can be defined as such. Moreover, there may be a discrepancy between the moment when patients realize they have no other chance of cure and the moment when this reality becomes clear to doctors due to their expertise. Consequently, it may be inappropriate to wait until patients enter the terminal stage of their disease to talk about RTD options.

7.2.3 A Burdensome Background: Considering Possible Solutions

Given these challenges, it seems that there is no ideal point at which to schedule an RTD discussion. Whether a patient has been newly diagnosed, is at an advanced stage, or has exhausted all available treatment options, this topic can be terrifying. Communication barriers may arise due to the difficult way in which death might be perceived both by patients and by healthcare providers, as well as because of the dangerous loss of hope patients may experience, a feeling that can impact their quality of life no matter the stage of their disease.

The best long-term option could, therefore, be to inform and promote information about RTD through specific programs and brochures[5] aimed at the wider public, rather than only targeting cancer patients. Given that cancer is one of the most globally widespread diseases and that most people have at least one close relative or friend affected by this pathology, RTD is a topic relevant to everyone, both as individuals and as members of a community of friends and relatives. Such a large-scale strategy of information and promotion would help to educate people about RTD and its value as a procedure, and it would increase the public perception of RTD as a common practice. In this way, the topic of RTD would no longer be a burden on the shoulders only of cancer patients. Brochures dedicated to RTD should be made available in cancer centers and in oncology private practices for cancer patients directly interested in the topic so that they can access adequate information. However, this information should also be disseminated to the general public in order to create awareness of the topic even among those not directly affected by it. Promoting wider knowledge of RTD would help to ensure that cancer patients do not feel alone. Indeed, if a large proportion of the population knows about RTD, then, when the time comes for patients to evaluate the option and make a decision, they would know they can talk easily with friends and relatives about it without having to explain to them what RTD is.

---

[5] A sample RTD brochure is available in the Appendix.

This could help patients overcome communication barriers, and, by discussing this option with people who already know what RTD entails, the patients could better and more easily evaluate RTD as an option.

Against the backdrop of dissemination of information about RTD to the general population, an individual discussion of RTD must be dedicated to cancer patients. As argued above, this discussion should not happen at the moment of diagnosis, when there is too much at stake. Rather, the RTD option should be presented at a later point, yet ideally at a non-terminal stage of the illness, in order to have the time to discuss and make decisions when there are other options available. This solution would allow patients to discuss and evaluate the RTD option after having independently acquired general information about this procedure through brochures and flyers.

The best strategy is for the physician and clinical staff to identify on a case-by-case basis the moment when a patient seems ready to face this discussion. In this evaluation, the role of healthcare providers is crucial, because they are in touch with the patient and are likely to know not only the medical history but also the psychological and emotional state of the patient. Physicians and clinical staff are the stakeholders with the best chance of understanding when would be the most suitable time for RTD to be discussed with patients.

The identification of a correct timing for the first RTD discussion is difficult to establish. The suggested path entails a long-term goal of widespread education about the topic in order to inform not only cancer patients but also the wider public, so that the RTD procedure becomes more familiar and widely understood. Then, given the difficulty of initiating a discussion about RTD with patients, it is suggested to evaluate each individual case to determine when a patient would be ready to face the topic, possibly avoiding the first appointment with the oncologist as well as the moment in which the patient realizes their therapy will need to be suspended and palliative care strengthened. Once a possible path for timing has been proposed, the next step is to identify the person or persons responsible for presenting and discussing the RTD option with the patient.

7.3   Step II—Choosing the Right Spokesperson

Conducting an individual RTD discussion with a cancer patient is a delicate task. The donation literature indicates that, when discussing donation with families, the type of healthcare provider significantly impacts consent outcomes (Kamal et al. 1997). Selecting the right spokesperson to take charge of presenting patients with the RTD program is a key step in the process, since communication may impact patient evaluation of the program and consequently the decision-making process.

Such a topic, given its difficult and potentially distressing nature, needs to be handled with extreme sensitivity and professionalism in order that patients do not perceive this procedure as a finality. Improvisation should be avoided in this type of communication. As a patient interviewed in a study conducted by McIntyre et al. (2013, p. 6) commented, this topic "has to be approached in a very delicate way, because you are asking for a tissue donation after death, it's kind of like indicating that you might not survive this".

In light of these considerations, although the physician is the individual who can identify the best time for communication, who should present cancer patients with the option to donate tissues after death?

7.3.1 The Treating Physician

The first option is the physician who oversees the patient's cancer therapy, and with whom the patient has built a relationship. This option has advantages and disadvantages. The main advantage is the relationship that already exists between patient and physician, because this may facilitate the discussion process. Patients would not have to deal with an unfamiliar person when discussing this delicate issue, and they would likely find it easier to ask questions and seek clarification. The physician has already developed a relationship with the patient: they meet on a regular basis and the physician knows exactly how the care plan is set up and is progressing. This might facilitate the physician in picking the right moment to talk to the patient, and to know how to conduct the discussion in a way that does not distress or overwhelm the patient.

At the same time, the established relationship between physicians and patients may also have some disadvantages in relation to discussion of RTD. By its nature, this relationship is not equally balanced. When patients are diagnosed with cancer, they experience stressful and overwhelming feelings. They are at the beginning of a new phase of life that will be filled with appointments and treatment in an environment that is unfamiliar to them. During this phase, patients may be weak and feel vulnerable because of their disease. Thus, patients regard physicians as professionals on whose experience and skills their life depends, which inevitably results in an unbalanced dynamic in which the patient is clinically and psychologically dependent on the doctor. Physicians possess medical and scientific expertise, and for this reason patients may rely on their opinion and evaluations when making treatment choices. Consequently, physicians are likely to have great influence on patients' evaluations and may condition their choices. By picking the physician as the recruitment spokesperson, a patient may consider RTD as an option encouraged by their physician, whose expectations the patient may wish to meet, even if they remain unsure and confused about this choice. If the physician personally presents the patient with the RTD option, the patient might feel under pressure to enter the program and may not want to disappoint the physician by refusing to participate. Consequently, patients could be reluctant to show uncertainty about an option offered by their physician.

Moreover, physicians may feel hesitant about bearing the burden of initiating this discussion with patients; they may find it too emotionally demanding, and they might also perceive this task to be one that falls outside their duties and role. In McIntyre et al.'s (2013, p. 6) study, a clinical staff member, when asked about discussing an RTD program with patients, responded: "I would be concerned approaching them [patients]; they are here for treatment, hope, survival, a cure, and here we are talking about what to do with their body afterwards."

7.3.2 *Ethics Consultants*

In order to avoid the disadvantages of the patient's physician being the recruitment spokesperson, the second option is to designate this role to a specifically trained RTD ethicist. As discussed in Chapter 6, RTD ethicists are trained to facilitate a decision-making process that supports the choices and preferences of the patient within the framework of RTD. Thus, such consultants should represent the most appropriate professionals to fulfill the task of discussing the unprecedented and delicate issues associated with this procedure.

RTD ethicists are consultants specifically trained to understand and engage with this program and, for this reason, they are not involved with the patient's treatment and care. This option offers the advantage of avoiding a patient's embarrassment if they do not wish to enter the RTD program, because no established relationship with related obligations exists between the patient and the ethics consultant. The pressure that patients may feel under with their physicians is less likely to arise in this scenario.

Although the absence of a pre-existing relationship can prevent the patient from feeling obliged to participate or from being too embarrassed to refuse, the option of an RTD ethicist as a spokesperson has disadvantages stemming from the lack of an existing relationship with the patient. The interaction with an outsider, in particular when introducing a topic, could create discomfort for the patient, given its delicacy and difficulty. Cancer patients may not welcome the intervention of a stranger presenting them with options for postmortem tissue donation. Any potential misinterpretation of RTD as a finality would be concrete.

The best strategy, therefore, would be to grasp the positive aspects of each professional profile so as to blend a synergic and combined process based on a multidisciplinary team. Presenting cancer patients with the option of RTD entails ethical controversies linked to communication that make it challenging to pick the right spokesperson. Physicians and RTD ethicists both present advantages and disadvantages in relation to this role. Given this, it may be best to break down the discussion of RTD into more phases that involve the cooperation of both the treating physician and the RTD ethicist. Such an approach would let patients become gradually familiar with the subject without feeling under pressure about their choice.

7.4 Phase Zero

As previously discussed, the long-term goal is to set up specific programs for disseminating information about RTD to the general public in order to increase awareness of the topic. With such a program in place, patients, at the time of their first discussion with the spokesperson, would have already autonomously obtained general information about RTD through literature available at hospitals or in private practices, or, at the very least, they would be familiar with RTD from hearsay. The goal to increase public awareness of the RTD process might be achieved through the distribution of brochures that are both informative and informal, and which are intended to promote the culture of donation in oncology research. This would help to break the ice in the first conversation of RTD. Nevertheless, effects of this process require time to be implemented and become concretely reliable.

7.5 Phase One

The first in-person presentation of RTD should occur when the treating physician believes it most suitable, it should be between the patient and the treating physician, and it should consist only of an explanation of what RTD is and how it works by presenting the content of the general brochure. Thus, the physician should maintain an informative role and individually explain what the RTD program offers and how it functions. The physician's approach should be absolutely neutral, and the conversation should be predominantly informative. The physician is more appropriate than any ethics consultant in this phase, because the physician has already established a relationship with the patient, and the patient is used to getting information from the physician. It would also avoid exposing cancer patients to a discussion with a stranger about what to do with their body after death, which might cause distress and a feeling of breach of confidentiality. In order to play this role in an RTD program, physicians should be adequately trained. As previously discussed, the manner in which the program is first presented to patients and families might significantly affect how it is perceived and evaluated within the framework of a decision-making process. This first conversation should remain on an informative level, and the physician should not at this time solicit responses or seek consent from patients concerning their willingness to enter the program. Due to their relationship with patients, physicians are best placed to inform rather than to collect responses.

At the beginning of the conversation, physicians should clearly state that they are not working within or for the RTD team; rather, they should emphasize that their task is purely an informative and communicative one. Physicians should explain (i) what RTD is; (ii) why RTD is scientifically relevant; and (iii) how RTD works. The physician should also make clear that withdrawal of consent is allowed at any time before death, and they should clarify that specific information concerning consent withdrawal would be further addressed upon patient request. Then, after presenting patients with the information about entering the program, physicians should provide patients with a brochure on the subject that summarizes everything that has been explained in person by the physician. The meeting should include time to deal with any clarifications sought by the patient. Physicians should ask patients to take their time when thinking about RTD as an option. They should then make clear that their personal involvement in the subject is terminated and that patients interested in the opportunity should contact the RTD ethicist, whose references and contact details would be included in the brochure. Physicians should remain available for any request for further clarification concerning the discussion.

7.6   Phase Two

The next phase concerns only those patients who report an interest in learning more about the RTD program through a dedicated RTD ethicist. This individual must have advanced interpersonal and counseling skills, as well as training in clinical ethics and RTD program-related decision-making. The RTD ethicist should be available at the hospital, but they should not have been directly involved with the patient's treatment and care up to this point. This lack of an existing relationship between the two figures is important because patients should not feel forced to participate in the program, and the absence of an existing relationship makes it easier for a patient to decide not to participate. It is important that patients rely on physicians for information and further questions about the subject, because the existing relationship should provide the patient with more confidence to seek clarification. At the same time, it is important that a person without any existing relationship with the patient be in charge of the next phase, since this individual will be seeking the patient's consent.

It is in this phase, which follows the physician's explanation of RTD, that the concrete recruitment takes place. Only interested patients should contact the RTD ethicist in charge of providing further information and collecting participants. It is important that the question about participation in the program is not asked by any other clinical staff member. The role of the RTD ethicist is crucial as it is at the heart of the patient's decision-making process. This figure has been appropriately trained and possesses the skills to provide patients with the specific kind of assistance and support required in the RTD discussion. It is therefore crucial that other figures do not overlap with the ethicist's work in this delicate phase. Once patients have been informed by the treating physician, it will be up to them to seek further information in order to decide whether to participate by contacting the dedicated RTD ethicist; they would be free to decline without having to provide any reason for this decision.

The prospective meeting with the RTD ethicist should include a verbal explanation of the whole process that emphasizes the aspects already discussed with the physician, and which adds further logistical details about the procedure. Such details should include the option of consent withdrawal at any time until the moment of death, as well as the options for consent withdrawal after death through close relatives; information on how collected samples may be linked back to the donor; the possible research uses of retrieved tissues; considerations on incidental findings; an indication that, despite the high value of tissues retrieved, there is no possibility of any new type of treatment being available for the donor patient; clarification that there is no remuneration of any kind available for the donor patient; and the timing and logistics of the cadaver's return to families. This second phase of information should be followed by a request to consent or to refuse to participate in the RTD program (IC2).[6] Any consent of the patient should be signed by a next of kin.[7]

---

[6]Samples of information sheets and informed consent forms are available in the Appendix.
[7]See Chapter 8.

7.7 Phase Three

If the patient gives their consent at the end of the second phase, further confirmation of this decision would be requested should the illness become terminal. Some patients who consent to RTD participation may recover, and thus would in effect no longer be participating in the program. Patients whose disease progresses, and who had previously consented, will enter the final phase of the consent process. In this phase, the patient will be reminded of all the points previously discussed in the earlier phases; this will be a prelude to seeking the patient's confirmation (or revocation) of their previous consent.

In this phase, patients will be presented with an informed consent form (IC3)[8] on which they can indicate confirmation or withdrawal of, or changes to, the consent previously given (IC2). IC3 will not be as detailed as the form that patients were presented with during phase 2 (IC2). In phase 3, patients will be provided with a copy of their previously filled and signed IC2 in order to read and further evaluate it, together with the new IC3 on which they have to declare that they have read the consent previously provided (IC2). In addition, they would have to check, by ticking the corresponding box, if preferences expressed in IC2 have changed.

If the preferences have remained unchanged, patients will have to sign IC3. If, however, the preferences have changed, patients will have to check the corresponding box on the IC3 form. If they have decided no longer to participate in the program, the IC2 will be destroyed in the presence of the patient (see consent withdrawal, Chapter 8 below). If, however, the patient has changed their mind only about some previously declared aspects, they will have to complete IC2 again in order to update their final consent form with all their preferences. Upon patients' consent, next of kin will have to sign to further allow retrieval of tissues.

The proposed articulation of informed consent in two stages is intended to allow prospective donors to reflect on their choices so that they have the chance to update or to confirm them should they become terminal. Although this structure gives patients room for further reflection, in certain situations it might pose barriers to the fulfillment of donors' wishes, such as in the event of an unexpected death that would prevent patients from confirming their wishes in IC3. Clinical staff and RTD ethicists are in charge of ensuring that patients have the chance—and the time—to sign the consent forms before their death. In the event that a patient encounters the option to enter the RTD program when they are already in a terminal stage, they will be offered the option to sign IC2 along with IC3.

However, if a patient unexpectedly dies before signing IC3, should a signed IC2 be considered valid? Imagine the case of a prospective donor strongly committed to research values and willing to contribute to an RTD program. After signing the IC2, they are progressing well with treatments, so there is no reason to offer them the chance to sign IC3. Yet the patient is involved in a car accident and dies unexpectedly before being presented with the option to confirm their willingness to donate tissues through IC3. Would the signed IC2 fulfill the requirements for tissue collection in this

---

[8] Samples of information sheets and informed consent forms are available in the Appendix.

example? My argument has strongly advocated the position that a patient's wishes should be honored after their death provided that they do not jeopardize the living. In this example, although the patient did not unfortunately have the chance to confirm their wishes, those wishes were expressed and formalized on the IC2 form and the wishes do not jeopardize the living, so in such an exceptional case the IC2 alone should be considered as meeting the requirement for donation. Any other conclusion would mean that the patient's wish to donate—reflecting their autonomous decision and informed consent—would be jeopardized due to the misfortune of premature death. In other words, IC3 is an important stage in the informed consent process, but in exceptional cases (such as the one discussed in this paragraph) IC2 can be considered valid as the formalization of informed consent.

The complex and delicate nature of RTD as a topic requires, in each phase, that the physician and the ethicist are especially careful in their use of language and choice of words when discussing RTD. Patients and families may be particularly sensitive to the words used by their interlocutor to express potentially distressing topics such as death and dying. To this end, in order to put patients and families at ease and not to make them feel uncomfortable in a particularly difficult moment, attention should be paid to tailoring these discussions to the audience. This is one of the specific skills in which the RTD ethicist is trained.

7.8 Vulnerable Populations and RTD

The main aim of this study is to describe an informed consent process, understood in the two meanings of the term, that is appropriate for RTD decisions made by competent adults. The choice of competent adults as a focal point is motivated by the fact that, within the framework of RTD—namely an altruistic decision as well as a free act without physical direct benefit to the patient—decision-making capacity is an essential element of the informed consent process. In other words, in the RTD context, the aim of acting in the best interests of the potential donor is hard to apply since the donor is no longer alive. Nevertheless, it is also important to briefly consider the issue of informed consent for RTD in relation to vulnerable populations, because patients with limited or no decision-making capacity, as well as minors, might shed interesting light on the general topic of informed consent for RTD.

As is the case in clinical research, the inclusion of patients with limited or no decision-making capacity raises thorny issues. Such concerns are explicated within the Nuremberg Code, which begins as follows:

The voluntary consent of the human subject is absolutely essential. This means that the person involved should have legal capacity to give consent; should be so situated as to be able to exercise free power of choice, without the intervention of any element of force, fraud, deceit, duress, overreaching, or other ulterior form of constraint or coercion; and should have sufficient knowledge and comprehension of the subject matter involved so as to enable him to make an understanding and enlightened decision. (Nuremberg Code 1949)

If literally interpreted, the Nuremberg Code would ban all research involving minors and adults with limited or no decision-making capacity. However, subsequent codes and guidelines[9] have clarified the position by explicitly allowing the inclusion of vulnerable populations within research provided that specific ethical conditions are met, such as that the research is in the best interests of the patient and that permission for participation is obtained from a legally authorized representative.

Enrolling in research participants who have limited or no decision-making capacity raises relevant concerns and has been subject to intense debate in research ethics that goes beyond the scope of this work. Nevertheless, in relation to RTD, it is important to know that the involvement of minors and adults with limited or no decision-making capacity within the framework of clinical research in the field of oncology is based on the patient's best interest standard. Cancer research participation by adults with limited or no decision-making capacity might allow for the availability of experimental treatments for those who have failed standard treatment options, and thus it might provide patients with the option of some direct benefit. This is also the case for minors, yet an additional consideration should be made for them. It is recognized that minors are a vulnerable population who should be included in oncology research with a specific focus on the development of new drugs or a combination of drugs. This need is grounded on the consideration that minors have particular physiological and clinical characteristics such that it is only partially safe and effective to consider valid the results obtained from oncology research on adults. This difference may impact the determination of potential benefits and risks, doses, age, appropriate administration regimens, assessment of potential drug interactions, pharmacokinetics, and pharmacodynamics. Given this, oncology research on pediatric patients might be needed because it makes available specific treatments for childhood cancers, which, from a biological point of view, are potentially different from those affecting adults (Aleksa and Koren 2002).

The world of oncology research provides an important framework of best practices to evaluate whether and how to enroll minors and adults with limited decision-making capacity. In addition, RTD poses unanticipated issues relating to donation. Indeed, the donation of postmortem samples, although representing a form of oncology research, has different characteristics compared to cancer clinical trials, so such donation requires further evaluation.

In the absence of a best interest standard linked to the option of RTD, and when no direct benefits are expected for those enrolled—as is the case for RTD—the participation of individuals with limited or no decision-making capacity raises ethical concerns. With regard to adult patients with limited or no decision-making capacity, further considerations should be made.

---

[9]Declaration of Helsinki (1964); Council for International Organizations of Medical Sciences (2002); Department of Health and Human Services, National Institutes of Health, and Office for Human Research Protections (2005). On this topic, see also Moreno (1997).

7.8.1 Adults with Limited Decision-Making Capacity

Decision-making capacity is not clear cut and its threshold can vary depending on the task the individual is asked to perform. It is a complex clinical phenomenon based on different variables, such as intellectual ability, cognitive skills, attention, concentration, ability to make probability determinations, ability to plan, and ability to solve problems (Rosenstein and Miller 2008). Within the clinical setting, patients may show impairment in specific performances while being able to make autonomous decisions in other contexts. Decision-making capacity can be pictured as a set of skills that "varies along a continuum from incapacitated to fully capacitated" (Rosenstein and Miller 2008, p. 439). Consequently, some patients may be unable to provide informed consent to enter a clinical trial, yet they might be able to pick a trusted individual with power of attorney. Such a spectrum of decision-making ability might also affect the context of RTD decisions. A cancer patient might have limited decision-making capacity—for example, the patient might be in the early stage of dementia—and thus have appointed a medical durable power of attorney or a legal representative,[10] but they might show interest in the RTD program. The opportunity to participate should be adequately assessed and thoroughly evaluated by the legal representative, because, in some cases, participation may represent a viable choice.

A separate case concerns adults currently without decision-making capacity but who, at an earlier point in life when they had decision-making capacity, had expressed interest in the donation of tissues. The ethical issues at stake in such circumstances are those relating to the impossibility of assessing the patient's position on the matter and to understanding whether the patient's wishes have changed since their earlier decision. The role of legal representatives is extremely delicate in RTD, but they are needed to make decisions on behalf of the patient. The thorough evaluation and resulting decision should be oriented toward the intent to act in a way that respects what the patient would have chosen had they had the opportunity to do so. In the case of tissue donation via RTD, assessment will need to be based on their experience or previous will, and the legal representative will be required to sign the consent form in their place.

The process for this kind of consent follows, with adequate modifications, the one proposed for competent adults. Clearly, for adults with little or no decision-making capacity, the legal representative plays a central role. With this in mind, phase zero occurs without particular changes. The brochures are made available for consultation by all patients, relatives, caregivers, and preferably, the general population. In phase one, on the other hand, the treating physician must engage in discussion with the legal representative first and then, if conditions permit, with the patient. Although seemingly simple, this step may not be trivial, as the views of legal representatives of adults with limited decision-making capacity might not coincide with those of caregivers taking them to medical appointments. Should the views of the caregiver who interacts with the treating physician not coincide with

---

[10]Legal representatives or individuals with medical durable power of attorney will be referred to in this book by the term "legal representative".

those of the legal representative, the treating physician should ask the caregiver for direct contact with the legal representative and schedule a meeting in order to present the opportunity of RTD donation. After presenting the legal representative with the basic information established for phase one, the physician should explain how to get further information about the program by introducing the figure of the RTD ethicist, whose contact details will be contained within the brochure. The legal representative should keep the brochure after the conversation in order to evaluate the proposal and reflect further.

In the event that the RTD program arouses the interest of the legal representative, they can schedule a meeting with the RTD ethicist and discuss further aspects. In the case of consent to enter the program, the legal representative will sign the dedicated form (IC2_LR).[11] As in the case of competent adults, the consent signed in the second phase must be confirmed or modified during phase three by the legal representative (IC3). Within this framework, the RTD ethicist plays a central role of mediation and support should conflicting positions arise between the legal representative and other relatives of the patient concerning RTD participation.

7.8.2 Minors

From an operational point of view, the informed consent process for minors follows, with necessary adjustments, that for competent adult patients. Parents or guardians, as well as the minors themselves, are central to the process. Minors represent, from both a social and a legal point of view, a sub-category of those who have limited decision-making capacity.[12] Indeed, their condition is inherent and unavoidable for all those in the early stages of human life. This condition, if not in particular situations of impairment, lasts until they reach the so-called age of majority, whose threshold varies according to national laws. Accordingly, only individuals who have legal empowerment and decision-making capacity can provide informed consent for minors. However, it should be noted that the literature has long been oriented toward an involvement of older minors and adolescents in the decision-making process by seeking, as far as possible, the "assent of the patient as well as the participation of the parents and the physicians" (AAP 1995, p. 316 n. 2). As minors develop, they "should gradually become the primary guardians of personal health and the primary partners in medical decision making, assuming responsibility from their partners" (Ibid.).

---

[11] Samples of information sheets and informed consent forms are available in the Appendix. In the main text, these are referred to as IC_2; IC2_LR; IC3; IC2_PG; IC2_M7-11; IC2_M12-17.

[12] As mentioned earlier in this book, this analysis is not focused on informed consent for minors. This paragraph is dedicated to their consent, but it is not intended as an exhaustive discussion of the topic.

Minors' involvement in the informed consent process, which is a critical issue both in care and in research,[13] is even more central when it comes to RTD. The decision whether to participate in the program should, as far as possible, take into account the personal inclinations and the will of the minor patient in order to make a decision that reflects their personal wishes. The values of RTD involve engagement, altruism, and participation. It is therefore vital for minors to feel part of the team and to be directly and actively involved in the informed consent process.

The responsibility for the final decision clearly lies with the parents or guardians, but in the field of RTD it is particularly relevant to do everything possible to ensure that the parents' or guardian's position is in agreement with that of the minor. Given this aim, the mediation and expertise of the RTD ethicist are of central importance.

Phase zero is common to all patients, including minors. Brochures and general information are made available to everyone to encourage knowledge about the topic. In phase one, the treating physician should find the most suitable moment to introduce RTD based on the development of the disease of the minor. This conversation should initially happen with parents/guardians, in the absence of the minor. Any conversation about RTD with the minor should take place only after the parents/guardians have given their consent to the minor being involved in the discussion. RTD conversation involves delicate topics and parents/guardians might not want their children to be involved at all. Such a desire should be respected.

Should parents/guardians and minors be interested in receiving further information about RTD, phase two begins. This takes place with the mediation and support of the RTD ethicist. If, upon discussion and evaluation, the minor and the parents/guardians intend to provide informed consent, it will be collected in two separate forms: an informed consent form dedicated to parents/guardians (IC2_PG) and an assent form dedicated to minors and formulated according to age groups, either 7–11 or 12–17 (IC2_M7-11 and IC2_M12-17). This precaution is so that the assent form is suited, as far as possible, to the cognitive abilities of the minor's age group, thereby making them feel as actively involved and engaged as possible within the process. In the case of minors close to the age of majority, it is important to present the patient with the consent for competent adults as soon as they have passed the threshold.

Consent signed within phase two should be confirmed, modified, or revoked in phase three, as in the case for competent adults. Again, the engagement of both parents/guardians and minors is central to this phase in which the final decision is made, and this cooperation should be ensured and encouraged by the ethicist.

The RTD ethicist plays a central mediating role between the two sides. Given the delicacy of RTD, it is not uncommon for conflicting opinions to arise between two parents or between parents/guardians and the minor regarding prospective participation in the program. Establishing rigid procedures to govern conflicts a priori could be counterproductive, and mediation by the ethicist, in case of conflict, represents an irreplaceable resource. However, it should be borne in mind that, given the

---

[13] Among literature on this topic, see Fleischman and Collogan (2008).

dramatic circumstance faced by patients who are evaluating whether or not to participate in the RTD program, it is always advisable to avoid family tensions arising from divergent positions, which can generate significant discomfort and stress, especially for the patient. In principle, it can be said that any participation of the minor in the program must be supported by both parents/guardians and the minor. When parents/guardians end up taking different positions concerning their child's participation in the program, even after the RTD ethicist's mediation, it is advisable not to proceed with RTD. When the conflict arises between parents/guardians on one side and minors on the other, thorough evaluations should be made. The final decision lies with the parents/guardians, but the position of the minor should be taken into account to a great extent. Given the RTD objectives of altruistic value and active engagement of the donor, parents' or guardians' consent to RTD is highly inadvisable if the minor objects. Conversely, if the minor wishes to participate and the parents oppose, the minor should not be allowed to participate. Given the already dramatic situation facing cancer patients, divergent positions and conflicts arising within the family from the possibility of participating in RTD should be avoided if possible. Hence, the RTD ethicist becomes, in this context, even more essential.

7.9 Informed Consent to Personal Data Processing

The informed consent process described within this chapter has focused on the development of an informed consent for this technique intended in two senses of the term: an informed consent in a *formal* sense, that is, an expression of a legally recognized dimension constituted for cancer patients who wish to donate tissues; and an informed consent in a *substantial* sense, whose requirements are modeled according to the autonomous authorization of the cancer patient.

However, when it comes to patients' consent to enroll in research, a clarification needs to be made. Within the clinical setting, and from an operational perspective, the informed consent expresses the willingness of the patient to undergo a specific research intervention. It witnesses the authorization of the patient to proceed, after having comprehended and competently evaluated available alternatives for the patient's condition. According to the *substantial* sense of informed consent, patients both assume responsibility for what they have authorized and transfer it to another's authority in order for it to be implemented. This meaning is separate but intrinsically bound to the institutional set of rules and policy governing this act, which the *formal* sense of informed consent refers to. This second sense involves a legally effective authorization from a patient and reflects the conformity to the social rules of consent that require medical professionals to seek valid consent from patients prior to performing medical or research interventions.

This form of consent must not be confused with the consent to personal data processing, which in Europe is governed by the recent Regulation (EU) 2016/679 on the protection of natural persons with regard to the processing of personal data. This kind of informed consent is concerned with managing the processing of patients' healthcare data collected within a trial and may involve special categories of personal data (sensitive data), such as genetic data. For this reason, this consent form is particularly relevant to the safeguarding of patients enrolled in research trials. Within this framework, in Europe, informed consent to personal data processing has a legal basis (Regulation [EU] 2016/679) according to which the processing of personal data must be lawful.[14] Thus, when patients enroll in research, they may be provided with both the informed consent form that witnesses their willingness to enter the trial and an informed consent form for processing of personal data collected within the trial.

Further analysis of the legal aspects and regulatory framework provided by the Regulation (EU) 2016/679 is beyond the scope of this study, which aims to consider general ethical issues relating to RTD rather than to discuss them from the perspective of a specific legal framework. Nevertheless, when it comes to considering the informed consent process for RTD, it is crucial to address issues pertaining to data processing that may apply to this form of postmortem research. From the RTD perspective, it might initially seem inappropriate to discuss consent for the processing of personal data, given that this data will be collected after the patient's death. But it is rather fitting. Consent to personal data processing applies to further potential involvement of blood relatives for inheritable traits and to the possibility of incidental findings (see Chapter 8 below). Moreover, consent pertains to the government of information concerning the ends to which collected personal data may or may not be used.

Consent for data processing for prospective RTD patients is appropriate for several reasons. First, when it comes to special categories of personal data, such as genetic data, specific traits of prospective RTD patients may pertain not only to them but also to their blood relatives. Thus, it is crucial to discuss with patients the relevance of personal data processing associated with RTD and the potential implications for their living blood relatives, in order for them more fully to evaluate informed consent. Given the importance and the delicacy of this topic, communication between patients and families on collection of genetic data should be encouraged and facilitated by the RTD ethicist. Specific preferences should be expressed by the prospective RTD patient about the preferred management of incidental findings that may pose concerns to their living blood relatives.

Secondly, the principles underlying the need for an informed consent process to govern RTD involve the respect for the recently deceased and the consideration of them as a person who led a life filled with personal values and choices that should be honored, even after their death, when these preferences do not pose challenges to the living (see Chapter 4 above and its discussion of the notion of the "once-alive"). Therefore, preferences concerning the processing of personal data should be collected and respected, including in relation to the specific ends for which data may or may not be used.

---

[14]In Europe, informed consent to personal data processing is not the only legal basis that permits use of personal data. According to Regulation (EU) 2016/679, there are five other legal bases that admit personal data handling, which include processing to safeguard the vital interests of the data subject or of a third party, and processing to carry out an action performed in the public interest or as exercised by an authorized official.

Moreover, respect for preferences established within the informed consent process is the greatest expression of RTD's values and its accountability, and it is the foundation of the whole donation process (see Chapter 8). The intrinsic nature of RTD is particularly sensitive to the values of altruistic donation and of respect for those who donate biological material in order to benefit the collective heritage by investing in the potential of research. For this reason, issues pertaining to personal data processing from prospective RTD patients are relevant and should not be disregarded. At the same time, it should be kept in mind that patients consenting to RTD are living through critical moments of their lives and should not be overwhelmed with excessive amounts of complicated forms to read and evaluate.

Given these considerations, the proposed solution is to create a unique informed consent form that incorporates simple and thorough aspects of data processing alongside the informed consent that authorizes RTD. This solution allows patients to express their consent to RTD by checking, within the same form, their preferences concerning how they would like their personal data to be managed. As a result, they would not have to complete two separate informed consent forms. The rationale for this solution is also due to informed consent—intended as authorization to undergo research—acquiring a particular role within the RTD framework. Indeed, the kind of research to which patients are consenting applies to interventions that take place after the death of the patient. Thus, while the meaning of donation itself is remarkable and should be detailed within the informed consent process, as well as the decision to perform (or not) this act, technical aspects concerning the collection *itself* are described in the consent form only at a general level, as their technical description may not represent, from the patient's point of view, additional useful information that contributes to making an informed decision. In this way, a line is drawn between the informed consent process for RTD and that for standard oncology research. In order to govern consent to participate in such an unprecedented procedure, several areas covered by standard informed consent forms used for clinical trials are inappropriate within the RTD context, such as the risks and benefits associated with the experimental treatment, trial length, randomization, placebos, and trial calendars for drug administration.

The unprecedented nature of informed consent for RTD, its particular aims, and the specific physical and psychological conditions in which it takes place, represent the grounds on which to establish a form of informed consent that stems from the informed consent process used in standard oncology research but is then shaped according to the specific needs of RTD. Thus, the informed consent process for RTD presents, within the same act, different souls. It is an informed consent in a *formal* sense, intended as a legally recognized dimension available for cancer patients who wish to donate tissues after their death; and it is also an informed consent in a *substantial* sense whose requirements are modeled according to the autonomous authorization of the patient.

References

Achkar, T., et al. 2016. Metastatic breast cancer patients: Attitudes toward tissue donation for rapid autopsy. *Breast Cancer Research and Treatment* 155 (1): 159–164.

Aleksa, K., and G. Koren. 2002. Ethical issues in including pediatric cancer patients in drug development trials. *Pediatric Drugs* 4 (4): 257–265.

Alsop, K., et al. 2016. A community-based model of rapid autopsy in end-stage cancer patients. *Nature Biotechnology* 10 (34): 1010–1014.

American Academy of Paediatrics (AAP) Committee on Bioethics. 1995. Informed consent, parental permission, and assent in pediatric practice. *Pediatrics* 95 (2): 314–317.

Bryant, J., et al. 2015. Oncology patients overwhelmingly support tissue banking. *BMC Cancer* 15 (1): 413.

Council for International Organizations of Medical Sciences, in collaboration with the World Health Organization. (2002). International ethical guidelines for biomedical research involving human subjects. CIOMS and WHO. http://www.cioms.ch/frame_guidelines_nov_2002.htm.

D'Alessandro, A.M., et al. 2008. Understanding the antecedents of the acceptance of donation after cardiac death by healthcare professionals. *Critical Care Medicine* 36: 1075–1081.

Declaration of Helsinki. 1964. Ethical principles for medical research involving human subjects. World Medical Association. https://www.wma.net/policies-post/wma-declaration-of-helsinki-ethical-principles-for-medical-research-involving-human-subjects/.

Department of Health and Human Services, National Institutes of Health, and Office for Human Research Protections. 2005. The common rule, title 45 (Public welfare), code of federal regulations, Part 46 (Protection of human subjects). http://www.hhs.gov/ohrp/humansubjects/guidance/45cfr46.htm.

DuBois, J., and E. Anderson. 2006. Attitudes toward death criteria and organ donation among healthcare personnel and the general public. *Progress in Transplantation* 16 (1): 65–73.

Embuscado, E.E., et al. 2005. Immortalizing the complexity of cancer metastasis: Genetic features of lethal metastatic pancreatic cancer obtained from rapid autopsy. *Cancer Biology & Therapy* 4 (5): 548–554.

European Commission, Data protection in the EU. 2016. Regulation (EU) 2016/679 of the European parliament and of the Council of 27 April 2016 on the protection of natural persons with Regard to the processing of personal data and on the free movement of such data, and repealing directive 95/46/EC (General Data Protection Regulation). https://eur-lex.europa.eu/eli/reg/2016/679/oj.

Fernandez, C.V., et al. 2012. Recommendations for the return of research results to study participants and guardians: A report from the Children's Oncology Group. *Journal of Clinical Oncology* 30 (36): 4539–4573.

Fleischman A.R., and L.K. Collogan. 2008. Research with children. In *The Oxford Textbook of Clinical Ethics*, ed. E. J. Emanuel et al., 447–460. Oxford: Oxford University Press.

Galushko, M., et al. 2012. Challenges in end-of-life communication. *Current opinion in supportive and palliative care* 6 (3): 355–364.

Genotype-Tissue Expression (GTEx) Project, ELSI sub-study run by Siminoff et al. http://www.siminoffresearchgroup.org/research-studies/gtex/.

GTEx Consortium. 2013. The Genotype-Tissue (GTEx) project. *Nature Genetics* 45 (6): 580–585.

Hanto, D.W., et al. 2005. Family disagreement over organ donation. *Virtual Mentor* 7 (9): 581–583.

Horne, G., et al. 2012. Maintaining integrity in the face of death: A grounded theory to explain the perspectives of people affected by lung cancer about the expression of wishes for end of life care. *International Journal of Nursing Studies* 49 (6): 718–726.

Hulette, C.M., et al. 1997. Rapid brain autopsy: The Joseph and Kathleen Bryan Alzheimer's Disease Research Center experience. *Archives of Pathology and Laboratory Medicine* 121 (6): 615–618.

Hyde, M.K., and K.M. White. 2009. Communication prompts donation: Exploring the beliefs underlying registration and discussion of the organ donation decision. *British Journal of Health Psychology* 14 (3): 423–435.

Kamal, I.S., et al. 1997. Does it matter who requests necropsies? Prospective study of effect of clinical audit on rate of requests. *BMJ* 314 (7096): 1729.

Krueger, R.A., and M.A. Casey. 2015. *Focus groups: A practical guide for applied research.* Thousand Oaks, CA: Sage Publications.

Laposata, M. 2017. A new kind of autopsy for 21st century medicine. *Archives of Pathology and Laboratory Medicine* 141 (7): 887–888.

Lindell, K.O., et al. 2006. Lessons from our patients: Development of a warm autopsy program. *PLoS Med* 3 (7): e234.

McIntyre, J., et al. 2013. Stakeholder perceptions of thoracic rapid tissue donation: An exploratory study. *Social Science and Medicine* 99: 35–41.

McVearry-Kelso, C.M., et al. 2007. Palliative care consultation in the process of organ donation after cardiac death. *Journal of Palliative Medicine* 10 (1): 118–126.

Merchant, S.J., et al. 2008. Exploring the psychological effects of deceased organ donation on the families of the organ donors. *Clinical Transplantation* 22 (3): 341–347.

Moreno, J.D. 1997. Reassessing the influence of the Nuremberg Code on American medical ethics. *Journal of Contemporary Health Law and Policy* 13: 347–360.

Nuremberg Code. 1949. In: Trials of war criminals before the Nuremberg military tribunals under control Council Law No. 10. Volume 2. U.S. Government Printing Office 181–182. http://www. hhs.gov/ohrp/references/nurcode.htm.

Pentz, R.D., et al. 2003. Revisiting ethical guidelines for research with terminal wean and brain-dead participants. *Hastings Center Report* 33 (1): 20–26.

Pentz, R.D., et al. 2005. Ethics guidelines for research with the recently dead. *Nature Medicine* 11 (11): 1145–1149.

Quinn, G.P. et al. 2013. Altruism in terminal cancer patients and rapid tissue donation program: Does the theory apply? *Medicine, Health Care and Philosophy* 16 (4): 857–864.

Rodrigue, J.R., et al. 2006. Organ donation decision: Comparison of donor and nondonor families. *American Journal of Transplantation* 6 (1): 190–198.

Rosenstein, L.D., and F.G. Miller. 2008. Research involving those at risk for impaired decision making capacity. In *The Oxford textbook of research ethics*, ed. J.E. Emanuel et al., 437–445. Oxford: Oxford University Press.

Rubin, M.A., et al. 2000. Rapid ("warm") autopsy study for procurement of metastatic prostate cancer. *Clinic Cancer Research* 6 (3): 1038–1045.

Schabath, M.B., et al. 2013. Healthcare providers' knowledge and attitudes about rapid tissue donation (RTD): phase one of establishing a rapid tissue donation programme in thoracic oncology. *Journal of Medical Ethics* 40 (2): 139–142.

Shah, R.B., et al. 2004. Androgen-independent prostate cancer is a heterogeneous group of diseases lessons from a rapid autopsy program. *Cancer Research* 64 (24): 9209–9216.

Spunt, L., et al. 2012. The clinical, research, and social value of autopsy after any cancer death: a perspective from the Children's Oncology Group Soft Tissue Sarcoma Committee. *Cancer* 118 (12): 3002–3009.

Sque, M., et al. 2005. Organ donation: Key factors influencing families' decision-making. *Transplantation Proceedings* 37 (2): 543–546.

Stevens, M. 1998. Factors influencing decisions about donation of the brain for research purposes. *Age and Ageing* 27 (5): 623–629.

Stouder, D.B., et al. 2009. Family, friends, and faith: How organ donor families heal. *Progress in Transplantation* 19 (4): 358–361.

Thombs, J., et al. 2005. Recruiting donors for autopsy based cancer research. *Journal of Medical Ethics* 31 (6): 360–361.

Van Der Linden, A., et al. 2014. Post-mortem tissue biopsies obtained at minimally invasive autopsy: an RNA-quality analysis. *PLoS ONE* 9 (12): e115675.

Wilkinson, T.M. 2005. Individual and family consent to organ and tissue donation: Is the current position coherent? *Journal of Medical Ethics* 31 (10): 587–590.

Williams, M.A., et al. 2003. The physician's role in discussing organ donation with families. *Critical Care Medicine* 31 (5): 1568–1573.

Willis, D., H. Draper. 2012. To make the unusual usual: Is there an imperative to discuss organ donation with palliative care patients? *International Journal of Palliative Nursing* 18 (1): 5–7.

Wood, K.E. et al. 2004. Care of the potential organ donor. *New England Journal of Medicine* 351 (26): 2730–2739.

Chapter 8
Informed Consent for RTD: A Closer Look at Ethical Issues

Abstract Drawing on the structure of the informed consent process provided in Chapter 7, this chapter explores the ethical issues that arise from Rapid Tissue Donation (RTD) decision-making. Guided by altruism, RTD represents a special agreement between cancer patients and science, which is regulated by respect and responsibility. This framework of values underpinning RTD is crucial for the ethical implications associated with this technique to be understood and appreciated within their context. These implications pertain to discussion about consent withdrawal, to how collected samples will be identified, and to their possible uses in research. Appropriate considerations will be given to incidental findings and to how blood relatives might be affected by RTD patients' decisions. The absence of concrete chances of translation into new treatment available for the donor patient upon RTD consent will be explored, as will the importance of an adequate appreciation of differences between kinds of benefits that might be associated with an intervention, and with RTD in particular. Different forms of altruism that might drive a donation will also be considered, as this aspect is extremely relevant to RTD. Finally, compensation for RTD donation and operational aspects such as cadaver restitution to the family will be discussed. As in the previous chapters, the aim of this chapter is to analyze relevant issues relating to RTD, without focusing on specific legal frameworks. Ethical considerations raised within this book can then be adapted to and harmonized with legal dimensions in force.

8.1 A Multi-phased Informed Consent Process

Rapid Tissue Donation (RTD) involves a particular kind of research, but one that has ethical, clinical and logistical implications.

As the analysis in the previous chapters has shown, the RTD informative process has been broken down into several phases so that patients have more time to become familiar with the topic and to make a decision that reflects their personal wishes. This informative structure has been proposed as a way of avoiding possible influences on patients deriving from unbalanced relationships with healthcare providers, since such relationships may jeopardize the patient's autonomy. Thus, the goal of phase zero of the process is to ensure that patients (and those around them) have already autonomously obtained general information about RTD, or at least have become indirectly familiar with RTD by hearsay, well in advance of making any decisions about it. This would help to break the ice in the first conversation about RTD. Within this process, the use of informative brochures[1] or informative sheets is crucial.

The next phase, at the heart of which is the involvement of the physician and in-person discussion of the brochure, is concerned with discussing what RTD is and how it works. In this phase, physicians have an informative role, explaining to patients (or elaborating upon the patient's existing perceptions of) (i) what RTD is; (ii) why RTD is scientifically relevant; and (iii) how RTD works.

The conversation with the physician in phase one ends the communication and information process addressed to all patients indifferently. After this phase, only patients who express interest in and openness to receiving further information on the topic will be further involved. Central to this process is the RTD ethicist who addresses details such as the option of consent withdrawal at any time until the moment of death, as well as the wishes of next of kin (NoK) to withdraw consent after a patient's death; information on how collected samples may be linked back to the donor; possible research uses of retrieved tissues; considerations on incidental findings; explanation that, despite the high value of tissues retrieved, there is no possibility of any new type of treatment being available for the donor patient; clarification that there is no remuneration of any kind available for the donor patient; and the timing and logistics of the cadaver's return to families. This phase ends with a request to consent to or refuse the procedure. If the patient gives their consent at the end of the second phase, further confirmation of their decision will be requested later in the course of the illness, should this advance and eventually lead to death, as too will be the authorization provided by the NoK. These aspects will be addressed below in further detail.

8.2 RTD as an Agreement Between Patients and Science

RTD is a technique that poses unprecedented ethical challenges. Before a more detailed consideration of relevant aspects raised by the various stages of the RTD informed consent process, it is crucial to outline the framework of values within which the ethical implications of RTD will be analyzed in order to comprehend and appreciate their meaning in context.

Biobanking has a long-established culture of tissue sample donation, as well as great knowledge about the values and meanings governing this act within the whole biobanking process.[2] Clearly part of the RTD program's potential and value relies on expertise derived from biobanking culture. The relevance of such knowledge

---

[1] A sample RTD brochure is available in the Appendix.

[2] For the European Biobanking Research Infrastructure, see the BBMRI network, available at www. bbmri-eric.eu.

therefore contributes to RTD programs in many ways. The novelty and the specialty of RTD programs pose ethical issues that are also found within the framework of biobanking practices. Hence, biobanking expertise can help inform a framework of practices and values for the specific needs of RTD.[3]

Each RTD patient is an active, unique and irreplaceable part of the RTD program, and this is how donors should feel. RTD patients (whether actual or prospective), their relatives, the healthcare professionals assisting the patients, the RTD ethicists supporting them, the researchers collecting samples and analyzing them to find new cures, and those who open up and develop new research scenarios should all feel part of the same team to which they make their personal, inimitable, and indispensable contribution.

Biological sample donation through RTD configures a particular model of oncology research, one which profiles personalized diagnostic and therapeutic responses for future patients, and, at the same time, safeguards the rights and promotes the informed choices of RTD patients who contribute with their donation.

The informed consent process governing RTD should represent, and be perceived to represent by each involved party, an agreement between RTD patients and researchers. This agreement is mediated by healthcare professionals, and it is safeguarded by RTD ethicists. Such an agreement is a pact of responsibility and cooperation between all the involved parties, each of whom is irreplaceable for the development of the new scientific approaches that cancer research requires. Thus, the act of donation exemplified by RTD involves, as far as possible, an altruistic, free, and voluntary choice by donor patients who wish to contribute with their tissues to the fight against cancer. The deep engagement, guided by faith and altruism, of each donor patient, which also affects their relatives and families, must be matched by a profound commitment from those on the other side of the process. This other side includes researchers, scientists, and other professionals who will work, at some point in the process, with samples collected through RTD. It is imperative that the agreement between patients and researchers is lasting and indissoluble. Researchers and other professionals who have been or will be dealing with the retrieved samples should handle those samples with intrinsic respect and should honor the aims that guided their donation and collection.

The research process is long and winding, and it involves various actors who work at different centers scattered around the world. In no way should the profound and intrinsic meaning of the agreement between donors and science be lost along the way or over time. RTD programs require a solid commitment by researchers to guarantee accountable, reliable, and responsible use and handling of collected tissues. These programs involve openness in the process with regards to the tissue uses for which researchers are accountable and responsible. RTD also encourages a deep engagement by the involved parties, from the engagement by RTD patients in the fight against cancer, to the engagement of researchers and others who interact with the donated samples and, in doing so, are part of the ultimate meaning of that

---

[3]This study is not intended to contribute to debates about the longstanding practices of biobanking, which is considered here purely because of the insights it can provide when reflecting on RTD.

donation. This many-sided engagement must represent a solid and lasting bond to guarantee the persistence of the meaning of RTD over time and is crucial for RTD ethical implications to be appreciated within the appropriate context.

8.3   Consent Withdrawal

Consent withdrawal is an important aspect of the informed consent process and stems from the acknowledgement that informed consent should govern the RTD donation procedure. This is because, as previously discussed in relation to the once-alive, the preferences and wishes (if any) expressed by individuals should be taken into consideration even after their death, as long as these preferences do not jeopardize living individuals. Since a cancer patient's hypothetical refusal to donate tissues through RTD is not likely to jeopardize living beings, then an informed consent procedure for RTD is acceptable (see Chapter 4 above). Again, a comparison with organ donation might help to underline this reasoning. Given the consistent direct benefit that organ donation provides to society (one organ saves one life), a hypothetical consent withdrawal from a prospective donor would represent a harm to the living who are on waiting lists.[4] This argument does not apply to RTD: a hypothetical refusal to donate tissues does not jeopardize anybody's life, because, once an RTD program is established, those who consent provide scientists with adequate material for research. Since any hypothetical refusal to donate to RTD is not likely to jeopardize the living, then, in principle, patients should be granted the right to withdraw their consent should they change their mind.

Discussion about consent withdrawal should be broken down into two steps. The first step entails the physician briefly clarifying, during the first conversation on RTD, that consent can be withdrawn at any moment before death (see Chapter 7 above). The second step, applicable only to patients interested in the topic and conducted by the RTD ethicist, involves a full discussion about consent withdrawal.

As with any form of oncology research, RTD allows a patient to withdraw their consent at any moment should they change their mind. In general oncology research, however, patients may withdraw their consent even after the protocol has started. Patients have the right to abandon the study at any time. Thus, patients who give their consent to participate in an oncology research protocol are able to commence their involvement in the study knowing that they always have the option to withdraw at any time. This condition does not apply to RTD, since the protocol takes place after a patient's death. This means that a patient can personally withdraw only until they die—in other words, their option to withdraw can only be exercised prior to the procedure and ensuing research taking place. Discussion with patients and brochures should be structured in an appropriate way that clarifies this feature of RTD while providing them with, as far as possible, the detailed information and reassurance necessary for making an informed decision.

---

[4]This example serves to emphasize features of RTD rather than to argue that organ donation should be compulsory; the latter debate is beyond the scope of this work.

8.4   The Role of Next of Kin in Consent Withdrawal

Should relatives be able to withdraw a patient's consent after the patient's death or to veto that consent? Do families have the right to object to the removal of tissues, even though the dead person gave their consent? Relying once again on the comparison with organ donation, it should be noted that, while organ donation allows no option of consent withdrawal by relatives after the donor's death—it would be bizarre and highly unacceptable for the deceased's family to ask for an organ back after transplant—this possibility might be foreseen for RTD donation. Undoubtedly, if the tissues had already been used in experiments or already inoculated in xenopatients, it could be problematic (and some research protocols protect themselves against this occurrence), but it would certainly be feasible, at least in principle.

From a legal point of view, NoK may have this prerogative, and many countries admit their right to veto. According to Lindell et al. (2006, p. 954),[5] "the patient's significant other needs to carry out the patient's wishes in order for the warm autopsy to take place. If the significant other decides not to pursue the patient's wishes, the warm autopsy does not occur." The option for NoK to veto that exists in some legal frameworks highlights the importance of shared decision-making among patients and their family. Patients should discuss their choices about RTD participation with their relatives, and they should inform them about their intended course of action. Sharing and discussing intended choices—with the facilitation of the RTD ethicist— would help to give relatives time to understand and make sense of the patient's choices. This sharing environment would also help patients face decisions made at the end of their life by having the support of their family and not to let unpleasant controversies arise after the death of the patient concerning their wishes. It might also be hoped that a greater public awareness of and familiarity with RTD—as proposed by the dissemination of information via leaflets and brochures (outlined earlier in this book)—would reduce the chance of conflicts arising after a patient's death.

Although shared decision-making in RTD is vital, it is worth noting that this dimension of a patient sharing their decision with their family is not trivial. As in oncology research, opinions may differ between the family and the patient concerning participation in research studies. In oncology research, situations sometimes arise in which relatives push their loved ones to undergo further experimental treatments in order to leave no stone unturned. This attitude is usually intended to encourage the patient not to quit fighting the disease and to keep hope alive. This inclination may differ within the scenario of RTD: relatives may be more likely to oppose a patient's wish to donate, because families do not want their loved ones to undergo unnecessary—since there is no direct benefit in return—yet invasive surgical procedures after death. It is worth noting that patients consenting to RTD know that their body will be taken away from their families very shortly after death in order for the retrieval to occur within the necessary window. This aspect may worry patients, since they know that their decision to participate in an RTD program may inflict additional suffering on their family who will temporarily be deprived of their loved one's body.

---

[5]See also Boyle et al. (2020).

Consequently, conflicting interests that may jeopardize patients' wishes can arise in this scenario.

Ideally, patients' decisions should be shared with their families in order to resolve conflicts before providing consent or refusal. In this process, the RTD ethicist has an important responsibility to provide information that can clarify the entire retrieval procedure. Despite the mediation of the RTD ethicist, it may happen that families oppose a patient's consent after the patient's death. In many countries, the authorization of the NoK is a requirement to proceed with postmortem retrieval. Yet, even in legal frameworks that do not envision this requirement, it should be noted that taking bodies away from families shortly after death and against the family's wishes can be a source of great distress that might exacerbate the already difficult moment of bereavement within the family. In fact, when the same scenario occurs in the organ donation framework—that is, when families oppose consent previously given by their loved ones—medical staff generally refuse to proceed with the explant, even if this would mean overriding a patient's personal wishes and renouncing the possibility of saving lives from the donation. In addition, medical practice generally allows families a veto on the use of bodies, even after a patient has consented to donate (Wilkinson 2005). The most cited reason for giving families the possibility to make the final decision is to avoid distress during the already difficult experience of bereavement (McIntyre et al. 2013).

It may also happen that families request tissue retrieval after a patient's death in the absence of a patient's consent. This circumstance may be less frequent, but it is significantly tricky. From an ethical point of view, if the patient had been given the opportunity to participate in the program and had explicitly refused, then any consent from the relatives should not be considered, since it would clearly oppose the expressed wishes of the patient. If the patient had not been made aware of the possibility of donating the tissues, the family can give consent on behalf of the patient if they believe that this would have been the wish of the deceased.

Also worth considering is the possibility that the NoK revokes consent after tissues have been retrieved and used for research aims. As previously discussed (see Chapter 1 above), tissues retrieved through RTD may serve to establish patient-derived cell lines and xenografts from primary and metastatic sites, which are recognized as the most appropriate models to study response and resistance to treatments. In the oncology field, patient-derived cell lines are frequently grown in the laboratory and inoculated with the same tumor as that from which the patient suffered in order to study its composition and analyze possible treatment options. Patient-derived cell lines are generally used in combination with xenografts. In this model, human tumor cells are transplanted, either under the skin or into the organ type in which the tumor originated, commonly using immunocompromised mice that do not reject human cells. This technique allows researchers to recreate a living environment in which the tumor may grow so that its development and corresponding treatments can be studied and understood.

The establishment of cell lines—and, mostly, xenografts—by using retrieved tissues with a patient's consent may result in issues relating to a potential consent withdrawal by relatives. In this situation, tissues, once collected, are used to create cell lines to be implanted. This means that their use is widespread and becomes an integral part of complicated and costly experiments, of ongoing research, and of published works. Therefore, should relatives seek consent withdrawal—especially a long time after tissue collection—it would become complicated to guarantee the absolute cessation of the use.[6] Different legal frameworks involve different paths associated with this circumstance: for example, consent withdrawal may be interpreted as not retroactive, meaning that tissues already used for research will not be deleted, but allowing for consent to be withdrawn for further use of tissues. However, this depends on different policies.

Consequently, consent withdrawal by NoK—amounting, in effect, to a veto on a patient's consent—is enabled by many jurisdictions, and the informed consent forms I have provided in this book adhere with the prevailing position to recognize this. This legal position is certainly understandable, since it aims to avoid further distress for the grieving family. Moreover, given the short time available for retrieval, it is important for patients and families to reach a shared vision concerning participation in order that unpleasant situations do not arise. Nevertheless, it is worth further inquiring, particularly in light of my analysis and argument in the preceding chapters, whether, from an ethical point of view,[7] relatives should have the right to override a patient's wishes. In other words, should NoK have the right to revoke the consent previously given before retrieval—and hence to deny and override a donor's wishes with respect to the donor's body, without that donor's knowledge or ability to respond since they are dead?

In Chapter 4, I argued that a new perspective should be adopted that involves shifting away from regarding those who have passed away as being "dead" and toward viewing them as the "once-alive". The aim of this perspective is to highlight, rather than to forget, the past condition of life and to bridge it to their actual condition of death. Moreover, it captures the way in which relatives regard their deceased loved

---

[6]To this end, the new European regulation GDPR and the closely related "Guidelines on transparency under Regulation 2016/679" developed by the "Article 29 Working Party" (WP29) can help. On the basis of GDPR and the WP29, it seems that, if consent is withdrawn, all data processing operations that were based on consent and occurred before the withdrawal of consent remain lawful. Nevertheless, the controller must stop the processing actions concerned. If there is no other lawful basis justifying the processing (e.g., further storage) of the data, they should be deleted by the controller. This issue brings up the much-debated discussions over prospective research uses of tissues with related implications for privacy, and over the difficulty to determine ex ante, at the moment of consent, the aims of using these tissues. The use of pseudonymization plays a key role in the process. However, a deeper analysis of the current GDPR implementation goes beyond the scope of this work. See European Commission, Justice and Consumers, Article 29 Data Protection Working Party "Guidelines on Transparency under Regulation 2016/679", available at: https://ec.europa.eu/newsroom/article29/item-detail.cfm?item_id=622227; European Commission, Data protection in the EU, "Regulation (EU) 2016/679 of the European parliament and of the Council of 27 April 2016 on the protection of natural persons with regard to the processing of personal data and on the free movement of such data, and repealing Directive 95/46/EC (General Data Protection Regulation)", art. 17(1)(b) and (3).

[7]Most legal frameworks recognize this right as belonging to NoK. The discussion here is intended to analyze the issue from an ethical perspective and to encourage further reflection and debate on whether the prevailing legal situation is based on a sound ethical analysis of the question.

one: not simply as a cadaver (or an abandoned vessel), but rather as someone who was once alive, whose once-aliveness remains meaningful in the minds of their loved ones, and whose life should be remembered and honored. If the deceased are thought of as the once-alive, their wishes, projects, desires, values, and preferences remain meaningful, and, as I have argued, those wishes should be honored provided they do not constitute jeopardy to the living. If this ethical argument is correct, then it would normally be wrong to allow NoK a veto over the once-alive's wishes. If the NoK have the right to override a donor's wishes, then it means that the dimension of once-aliveness is being denied, resulting in the deceased person becoming merely a cadaver over whom others make decisions.

It is certainly the case that denying NoK the right to withdraw donors' consent to RTD might in some cases cause stress and discomfort for those who are experiencing grief.[8] Relatives might oppose tissue collection and might feel disrespected and outraged by having the body of their loved one carried away immediately after death to enable tissue collection. It is to be hoped that such unpleasant situations might be avoided, or at least mitigated, through mediation and encouragement of a shared decision-making between patients and relatives by relying on the professional assistance of an RTD ethicist; in other words, the more robust the informed consent process, the less likely it is that conflicts will arise. Nevertheless, successful shared decision-making is not guaranteed, there may be occasions when conflict arises after a donor's death, and the pain of grief may impact on the NoK's view about RTD. In these circumstances, painful though it is, a donor's right to make autonomous decisions pertaining to tissue donation should be honored after their death by those who are alive. This respect not only honors the wishes and values of the once-alive, but it also protects what I have argued to be important ethical principles: the individual's right to autonomous decision-making, and their right to have their wishes honored provided they do not jeopardize the living.[9]

Nevertheless, the denial to NoK of the chance to withdraw consent to tissue collection is potentially painful. It is to be hoped that the more people become acquainted with RTD as a valuable procedure offered to cancer patients—through the implementation of RTD programs, brochures, and communication targeted at the general public (see Chapter 7 above)—the less likely it will be that NoK will be distressed by tissue collection, and the more likely it will be that they might rather appreciate the potential and value of RTD. Above all, it is worth considering the ethical arguments for and against an NoK veto of a donor's informed consent, and to reflect on whether such a veto strengthens or weakens the RTD informed consent process.

---

[8]While stress and discomfort, in particular for those experiencing grief, are undoubtedly overwhelming, they do not (usually) involve jeopardy as intended within this analysis, which is rather associated with a risk of death or serious harm.

[9]After all, this is not the only circumstance in which those who are alive are expected to respect and honor the decisions made by the deceased. For example, NoK do not normally have a veto over wills, nor are they allowed to oppose postmortem procedures, such as autopsies, that have been ordered by the authorities.

8.5 How Can Retrieved Samples Be Traced Back to Donors?

Particularly relevant for consent withdrawal is knowing how collected samples are identified. Prospective RTD patients should be informed about how their samples will be identified and possibly linked back to the identity of the donor. This issue has important repercussions for the privacy of blood relatives because data retrieved from collected samples might reveal sensitive healthcare information whose circulation should be limited. Biobank culture in Europe has long-established protocols to manage identification and anonymization of donated samples in accordance with ethical standards and normative frameworks, with a specific focus on the GDPR, whose analysis goes beyond the scope of this work. Nevertheless, it is worth mentioning that the most common method currently used is called pseudonymization, which is the

> processing of personal data in such a manner that the personal data can no longer be attributed to a specific data subject without the use of additional information, provided that such additional information is kept separately and is subject to technical and organisational measures to ensure that the personal data are not attributed to an identified or identifiable natural person. (GDPR, art. 4[5]5)

Regardless of the method used by the RTD program, prospective RTD patients and their families should be informed about how retrieved samples might be re-identified in order to make an informed decision.

8.6 Possible Uses in Research of Tissues

One of the ethical issues raised by RTD relates to possible uses of tissues in research. This debate regards the right (if there is one) of patients to be informed (and the extent of that information) about future uses of their tissues in research. In other words, it is important to consider whether to inform patients at the time of consent about the possible uses of their tissues, and, if so, how much information should be provided.[10]

Drawing on the comparison with organ donation, it should be noted that, whereas organ donation has a well-defined purpose and does not allow for further donor preferences (it is not possible to express preferences about the recipient), the various options relating to tissue use could, in theory at least, admit the possibility of the donor expressing a preference regarding the field in which the tissue is employed. The donor may in fact be allowed to authorize the use of donated tissues only in limited areas of research, such as oncology.

---

[10] Again, some legal frameworks have clear policies to govern this issue. The analysis here leans toward an ethical dimension, and the solution proposed—abstract from specific legal codes—allows patients' rights to be respected while also fostering scientific progress.

Should patients consenting to donate their tissues for research receive additional details regarding the kind of research their tissues will be used for? Regardless of whether this information is necessary or not, it should be kept in mind that determining ex ante, at the moment of consent, what the future uses of tissues might be is a challenging task. Scientific advances are unpredictable, and tissues collected today could become a precious future resource for areas that are not even imagined today.

On the one hand, it has been argued that donors should be made aware of possible future uses of their donated tissues, since this is their right. To this end, a specific form of informed consent has been proposed to govern donation by living patients of tissues for research aims, generally in relation to biobanks. This is known as "dynamic consent", and it constitutes an approach to consent that engages individuals in decisions about how their personal information should be used. It involves a personalized, digital communication interface that enables two-way communication between participants and researchers, and that puts participants at the center of decision-making (Kaye et al. 2015). This form of consent allows patients to constantly receive notifications about the studies that their donated tissues could be used for, so that they can choose which ones to authorize. Although interesting, this kind of informed consent may not represent a solution relevant to RTD, because RTD donors are deceased and hence no interaction and preference update can occur.

On the other hand, since scientific developments are unpredictable, establishing too many constraints on allowed uses of tissues through informed consent would represent a barrier to scientific progress. The value of donation entails an agreement between the donor and the RTD program, so it can be argued that donors should have the right to know, in broad terms, the purposes for which the donated tissues will be used without this implying the right to have a negative impact on scientific development. Much hinges on how extensively "in broad terms" should be defined. In general, some cancer patients who donate tissues through RTD may be relieved to know that their biological material will be used for oncology research development and not for something such as cosmetics. After all, part of the motivation of RTD donors—as will be discussed later in this chapter—may be the altruistic attempt to help those affected by the same pathology from which they are suffering. The knowledge that donated tissues will be used to help someone in the same situation may be a driver to donation and should thus be respected and encouraged.[11] Yet, for other patients, acknowledging whether their tissues will be used in oncology research or in other areas may pose no relevant concern at all. The different positions of donors should be adequately taken into account and these preferences should be, as far as possible, reconciled with the needs of scientific research in order to foster its development. Thus, both to respect donors' preferences and to encourage scientific innovation at the same time, an operative solution could take the form of grouping within the informed consent form prospective areas of research that collected tissues might be used for. This would allow donors to express their preferences on a multi-option form that asks them to check the boxes that correspond to the research area for which they are willing to consent to their tissues being used. Giving donors the option

---

[11] For European regulations on this topic, see the GDPR.

to choose which kind of research areas their tissues are used for might strengthen the concrete value of donation and its grounding in altruism. In addition, this option substantiates the bond of engagement and participation between patients and society. The empowerment of such a bond is particularly relevant within the critical moment in which donor patients are living and may represent for them a way to find a positive aspect to the dramatic evolution of their disease.

From an operative point of view, prospective areas of research should be included within the informed consent form and properly discussed with potential donors. This inclusion should be adequately balanced so that patients' preferences are valued without creating barriers to research development, while also ensuring that patients are not overloaded by excessively detailed and technical information. The manner in which research areas are grouped as options within the informed consent form should clearly reflect these aims. To avoid overwhelming patients, research areas should not be broken down into an excessive number of groups; at the same time, the groupings should be consistent and clear so that patients are able to understand what they can consent to. Consequently, the proper grouping of research areas on the consent form is not a trivial matter.

One group needs to be dedicated explicitly to oncology research. In oncology, prospective RTD patients may tend to feel a strong bond with other patients in their condition (Amir and Haskell 1997; Barr 2006; Quinn et al. 2013; Sanner 2006). For this reason, when deciding to donate tissues for research, patients may feel relieved to know that their action will, in some way, help other patients affected by the same disease. There is a strong sense of identity and altruism between members of the same group, and donors should be able to decide that their tissues are used for research in this area. At the same time, oncology is a broad research area, whose boundaries are wide enough to allow researchers to extensively carry out their studies.

Given the strong sense of altruism that grows between members of the same group, prospective donors might express the wish to have their tissues used exclusively in research aimed at finding treatments for the same kind of tumor they are affected by. Although this might seem a justifiable reason for preference, this request should not be embraced. Oncology is an umbrella for an extensive number of different tumors that require personalized and different approaches. Splitting the research area into different branches would pose problematic barriers to further development of treatments and may prevent scientists from gathering crucial aspects of tumor onset and evolution.

A second group might represent research aimed at studying the onset, evolution, and treatments to fight pathologies other than oncology. This area is extremely wide, but it is best not to divide the different branches of medicine into too many groups. This solution would avoid an excessively long consent form being presented to prospective donors, while at the same time ensuring that barriers to research development are not created.

A third group might be dedicated to the development of products like preventive tools (vaccines), diagnostic kits, or other commercial uses of the sample, and a fourth group should be dedicated to non-medical areas of research, such as cosmetics. It should also be taken into account that many future uses might be unknown, and this should be clearly mentioned. Thus, a fifth group should be dedicated to unknown purposes that do not fall under any of the other groups.

The resulting architecture of the informed consent form allows prospective RTD patients to consent to or refuse the use in different research areas of their samples, according to their preferences, by selecting the appropriate boxes. This has the great advantage of not letting patients reject RTD because of the fear of having their samples used for purposes that, for whatever reason, they consider to be inappropriate. The multiple options presented on this informed consent form allow every patient to exclude research areas to which they are not willing to contribute, without preventing them from the opportunity of participating in the donation process.

The five groups of research areas are intended to represent simply a concrete and operative—and non-exhaustive—example of how the development of scientific research may be encouraged and fostered while simultaneously safeguarding the preferences and values of donor patients. One of the greatest limits of this solution is that it might be difficult to trace the boundaries of research areas, as some of them might overlap. With this in mind, the scenario presented above is a possible solution, but not necessarily a definitive one.

When it comes to selecting the future uses of tissues, discussion with the RTD ethicist is particularly relevant as it might help patients to understand the different kinds of research and facilitate the accommodation of patients' and families' needs and values. A specific point that the RTD ethicist needs to address in this delicate conversation concerns which parts of the body may be affected by different research areas. For example, the wish for the bodily integrity of the deceased, such as to allow an open casket funeral, can be extremely meaningful for some patients and families who might experience pain upon seeing the potential disfigurement caused by significant tissue removal from certain areas of the body. Cosmesis, for example, generally involves retrieval from the face, and this should be highlighted by the RTD ethicist within a conversation that considers the religious, cultural, and personal values of patients and families. As highlighted later in this chapter, the "replenishment" of the body after retrieval of tissues should be granted with no additional expenses incurred by the family, yet patients and families should be made aware of tissue retrieval in parts of the body that may be particularly meaningful.

8.7   Incidental Findings

Recent advances and profound changes in biomedical research raise the ethical question about the adequate management of so-called incidental findings.[12] Incidental

---

[12]This section considers incidental findings specifically within RTD. For examples of the general literature on incidental findings, see: Berkman and Hull (2014), Bookman et al. (2006), Cadigan et al. (2011), Croyle and Lerman (1999), Gray et al. (2016), Green et al. (2013), Kahneman and Tversky (1979), Kalia et al. (2017), Kirschen et al. (2006), Kleiderman et al. (2013), Mandava et al. (2015), Meacham et al. (2010), Murphy et al. (2008), Presidential Commission for the Study of Bioethical Issues (2013), Ravitsky and Wilfond (2006), Robertson and Savelescu (2001), Royal College of Radiologists (2011), Shalowitz and Miller (2005), Wolf et al. (2008).

findings are previously undiagnosed conditions that are discovered unintentionally—in other words, their discovery was not the primary or secondary purpose of tests, analysis, and other practices conducted in clinical research or medical care. Given the unintentional nature of the discovery, which may be known only a long time after the collection of the patient's information, the government of acquired information poses ethical quandaries that have specific relevance within the field of general research, and especially in postmortem research.[13] The question is whether, and to what extent, scientists have duties to analyze and return to patients (or research participants) incidental findings generated in research with a specific focus on genetic data intended as "personal data relating to the inherited or acquired genetic characteristics of a natural person which result from the analysis of a biological sample from the natural person in question" (Regulation [EU] 2016/679).

Consider the following two examples. In the first example, a Phase 2 oncology trial tests a new combination of drugs for bladder cancer. While analyzing a blood sample retrieved for research aims within this trial months earlier, scientists discover a genetic mutation of a patient for a specific condition. In the second example, a patient consents to donate to oncology research some biological material left over from his biopsy. Years later, when analyzing the sample within a research project aimed at developing new treatments for gastrointestinal cancer, scientists detect a *BRCA* mutation. Should these patients be provided with information concerning their susceptibility to develop diagnosed or suspected conditions?[14] What if the data that revealed susceptibility to a disease had been collected a long time previously? And what if, at the time the incidental finding arises, no treatment had been discovered to control the condition at stake? These questions represent only some of the ethical quandaries raised by incidental findings.

The adequate management of incidental findings in research may be analyzed according to the framework provided by the basic principles of autonomy, beneficence/non-maleficence, and justice. From the perspective offered by the principle of autonomy, the participant's right to decide to be made aware of incidental findings from the research pertaining to their health should be recognized and taken into account. This is particularly the case when such information reveals crucial conditions that need adequate and timely treatment.

Although the primary aim of biomedical research is to increase what has been called generalizable knowledge, respecting participants enrolled in research is a requirement. According to ethical regulations, scientists are required to find a balance between the benefits and potential risks to which participants in a given research may be exposed. Research involving humans (and human samples) must be based on the principle of beneficence (maximization of benefit) and non-maleficence (minimization of harm). In some contexts, the disclosure of incidental findings can benefit

---

[13]Incidental findings occur in clinical practice too; however, this section is considering their relevance within research with a particular focus on the RTD framework.

[14]On this topic, see also: Andorno (2004), Chadwick (2004), Clift et al. (2015), Harris and Keywood (2001), Jelsig et al. (2015), Kaphingst et al. (2016), Laurie (1999), Ost (1984), Rhodes and Capitulo (2006), Shaw (1987), Townsend et al. (2012), Viberg et al. (2016), Wilson (2005).

the research participant, such as if the finding leads to the option to treat a condition before its development, thereby maximizing the chances of a cure. On the other hand, failure to report an incidental finding may cause potential damage to the research participant. When the condition is treatable and early treatment maximizes the chance of a cure, failure to disclose prevents the patient from taking action. Similarly, if the finding concerns a mutation that involves relevant implications for inheritable traits, not being informed may cause potential harm. Yet, beneficence and non-maleficence are not always achieved by disclosure. Sometimes, reporting an incidental finding may damage the patient, such as when the condition incidentally found is one with no available treatment options. Reporting an incidental finding might allow the relevant individual to prepare themselves psychologically and practically for the development of the disease, communicate with family, and engage in collective planning. However, it might also cause the patient stress and discomfort years before the onset of the disease, such as in the case of Huntington's disease. Therefore, the adequate application of beneficence and non-maleficence to incidental findings is not clear cut and needs a case-by-case evaluation for adequate balance.

With regards to incidental findings, the principle of justice involves not excluding disadvantaged social groups from the potential benefits associated with the communication of the findings. Incidental findings might reveal a potential condition that needs further analysis in order to obtain a validation of the results according to clinical standards, or even the use of genetic counseling services. The financial costs of the assessment of incidental findings should not present a barrier to disadvantaged groups of patients.

The right to know as applied to incidental findings relating to an individual's health is exemplified by the principle of autonomy. Precisely interpreted, however, this principle may lead to the opposite conclusion. The right to know, grounded in the principle of autonomy, might also become the right not to know. In other words, patients may express their right not to know information about their health. Communication of incidental findings can be interpreted as an ethical duty of the scientist based on the principle of beneficence and non-maleficence toward the participants in the research; however, there are relevant arguments in support of non-disclosure when the findings may not be a benefit for the participants. For example, this is the case in relation to findings with uncertain clinical validity or those that might induce potential overdiagnosis and overtreatment in addition to stress, uncertainty, and psychological harm.

The ethics of the right not to be informed about incidental findings leads to complex debates. This right has been officially recognized by the UNESCO Declaration of the Human Genome and Human Rights, in which it is stated that the "right of each individual to decide whether or not to be informed of the results of genetic examination and the resulting consequences should be respected" (UNESCO 1997, art. 5). However, the extent to which this right may be exercised needs further reflection, in particular when the implications of deciding not to know relevant information concerning health do not pertain exclusively to the individual who decides to exercise that right. Indeed, when genetic information is at stake, the right not to know might have relevant implications for blood relatives, so its application should be adequately evaluated. Thus, a limitation of the exercise of this right might be foreseen if that ignorance of personal healthcare information can cause harm to others. This is the case for incidental findings relating to genetic predispositions shared with blood relatives that could potentially lead to the development of diseases for which there is the chance of therapeutic or preventive measures. In these cases, patients' right not to know should be adequately balanced with the principle of beneficence toward their blood relatives who may themselves benefit from the discovery due to a genetic disposition to a condition.

Within the RTD process, incidental findings play a key role and raise ethical considerations that relate to the principles of autonomy, beneficence, non-maleficence, and justice. Investigations performed on tissue samples retrieved from cancer patients who consented to RTD may happen to show genetic mutations linked to the susceptibility to develop specific diseases. While the occurrence of incidental findings generally poses concerns about the prospective disclosure of retrieved information to the donor patient, the novelty of the RTD scenario requires these ethical concerns to be approached in a specific way.

Since tissue retrieval takes place after the death of the patient, the main concern within the RTD framework relates to incidental findings' implications for patients' blood relatives. Indeed, there can by definition be no concerns about any possible reporting to the donor of the tissues. Thus, the right not to know has to be viewed from a radically different perspective. In most research, any reporting of incidental findings may to some extent be hindered by the patient who originally owned the biological material exercising their right not to know; within the RTD framework, this possibility does not apply. Any disclosure within RTD regards the blood relatives of patients, and this feature of research on RTD-retrieved tissue needs to be carefully evaluated.

Although reporting of incidental findings is not targeted at the RTD patient, this is not to say that the patient might not be, at least emotionally, involved in this decision until death. In this situation, it is important for the RTD patient to take into account potential implications for blood relatives involved with consent to RTD.

Any reporting of incidental findings, whatever their long-term positive outcomes, always involves someone being made aware of aspects that immediately generate anxiety and stress about the unknown. Incidental findings can have entirely negative implications for those to whom they are reported. Given this, prospective RTD patients may be concerned about the possibility of imposing on blood relatives the positive or negative consequences deriving from what is conceived to be a personal choice of tissue donation. Such a concern might be exacerbated by the awareness that, when and if such potential findings are reported to blood relatives, the patient will already be dead and will be unable to support blood relatives in the management of information and in the decisions that will derive from it. These considerations acquire a dramatic relevance, as they might sharpen the feeling of guilt and responsibility that cancer patients may exhibit toward family members whose lives are already severely impacted by the consequences of the disease.

The evaluation of incidental findings is also impacted by support to patients and the level of agreement that family members and relatives have with regard to the option of RTD enrollment. When different opinions emerge among family members about the participation of a patient in the donation program, potential implications related to incidental findings may create additional concerns. Family members may sometimes express concerns about the possibility of their loved one being part of an RTD program. This hesitation is grounded in various subjective and objective reasons and may arise from the fear that the body of the loved one would undergo further invasive interventions after death. Therefore, prospective RTD patients might feel cautious about any incidental findings being reported to their families who did not even agree on participation in the program.

Therefore, the unprecedented framework of RTD can generate several concerns relating to the reporting of incidental findings.

The right to be (or not to be) informed about incidental findings acquires an immediate transitive value for blood relatives, as any potential information will arise at a time when the tissue donor is no longer alive. Indeed, the implications of such a transitive value demonstrate that blood relatives are central in this evaluation. Despite the undeniable role RTD donors play within this circumstance, the choice whether to be informed or not should incidental findings occur should involve those directly involved with potential implications, namely blood relatives.

The relevance of incidental findings and the delicacy of information potentially released to blood relatives should be adequately explained and further discussed with RTD patients and families in a conversation with the ethicist involved in the RTD process. Within this framework, blood relatives should be assisted in evaluating the potential burden in terms of psychological distress and healthcare outcomes in order to make an informed choice about incidental findings.

Blood relatives should be given the chance to decide whether potential incidental findings be reported to them. In this way, if cancer patients provide consent for RTD, blood relatives are always in a position to exercise, for themselves, the right to know or not to know should any information arise. Therefore, RTD patients' blood relatives should be given information sheets[15] that explain what incidental findings are and how they impact blood relatives of an RTD patient as well as a consent form to formalize their choice. Prospective consent to report incidental findings to blood relatives should be formalized in the informed consent form with the name, blood relationship with the RTD donor, and contact details should incidental findings arise.

Moreover, blood relatives of an RTD patient should be informed on how to withdraw from the reporting of incidental findings or to consent in case they refused in the first place by reporting their preference directly to the RTD program, and this should be clearly expressed on the information sheet they are provided with. This choice is intended to apply only to themselves and not to other relatives who might be affected by the finding.

The RTD program is in charge of contacting blood relatives of the patient in order to inform them of the option to be reported incidental findings. It is also recommended that incidental findings be reported by adequately trained professionals: the reporting of incidental findings involves dealing with delicate and sensitive issues from both

---

[15] A sample of an information sheet for blood relatives is available in the Appendix.

a clinical and an ethical point of view, so they should be handled by a professional with the appropriate expertise, skills, and training.

The transitive application of incidental findings' reporting within the RTD context emphasizes the potential implications in terms of beneficence and non-maleficence arising from disclosure to blood relatives. It is worth noting that not all incidental findings have the same relevance. They can be more or less reliable and accurate, they may need more or less testing for their assessment, they can relate to mild or very serious pathologies for which a treatment may or may not exist, they may concern pathologies that have repercussions for reproductive choices, and so on. These variables have a clear impact in terms of beneficence and non-maleficence for the people to whom this information is reported.

An accurate assessment of the type, accuracy, and reliability of incidental findings in order to evaluate potential disclosure acquires relevant significance from an ethical point of view. While there are divergent opinions on the topic, the prevailing view is to evaluate the best interest standard[16] for the individual to whom communication of the incidental findings is targeted. This standard reporting of incidental findings should be balanced according to three parameters: analytical validity, clinical actionability, and clinical relevance (Schaefer and Savulescu 2018). A result is analytically valid if it identifies a condition in an accurate and reliable way according to the standards normally adopted in the relevant research area. A result has clinical relevance if it reveals a risk that is both relevant and recognized for a significant medical condition, and/or if it presents implications for reproductive choices. Medical actionability refers to the association of the finding with the presence of a recognized therapeutic or preventive intervention, or with other possible actions capable of changing the clinical course of the disease or of the medical-reproductive condition; it also concerns whether the knowledge can significantly impact the projects and life plans of the subject involved.

The best interest standard when understood according to the reported parameters does not meet with univocal consensus. Among other things, the option to substantiate the roles of autonomy, privacy, and interests has been suggested in order to consider them as alternative standards to evaluate whether to report incidental findings (Schaefer and Savulescu 2018). However, whether to report incidental findings, and, if so, according to what conditions, are highly debated and controversial issues that lie beyond the scope of this book's focus on the RTD process.

Issues and considerations discussed here are not intended to prescribe what should be done or to enforce the best way to manage incidental findings. Each case raises different sets of expectations and values in individuals involved, requiring a specific analysis and balance to achieve responsible management of incidental findings. This section reflects a possible solution (but by no means the only feasible, or even the best, one) to govern incidental findings within the particular framework of RTD.

As has previously been discussed, it is important that prospective RTD patients and their families be informed about how collected samples will be (de)identified and

---

[16]See Bookman et al. (2006), Green et al. (2013), Hall et al. (2013), Ravitsky and Wilfond (2006), Wolf et al. (2008).

possibly linked back to the identity of the patient from whom they were retrieved. One reference technique in the field is pseudonymization, which is the processing of personal data in such a manner that they can no longer be attributed to a specific data subject without the use of additional information. What happens if incidental findings are detected within pseudonymized samples? In Europe, the answer is regulated by the GDPR framework. However, without entering into a legal discussion and a consideration of biobanking-related best practices, it is important to underline that incidental findings can be reported even if data are pseudonymized because the RTD program, which is responsible for their management, is able to re-identify the donor and activate (if consent had been given by the RTD patient) the incidental findings reporting process.

In the case of incidental findings that emerge during research practices conducted by secondary researchers, such as in research collaborations, studies, and multicenter research projects that do not correspond to the RTD program research group (primary researcher), it is the responsibility of this secondary researcher to contact the primary researcher responsible for the research project in which the personal data were originally collected, so that the latter can provide the communication of the incidental findings to the RTD participants' blood relatives, (Commissione per l'Etica della Ricerca e la Bioetica del CNR 2018). It is worth noting that the responsibility of the secondary researchers in contacting the primary researcher if an incidental finding occur should be clearly reported within the agreements or conventions that bind them.

8.8 Concrete Chance of Translation into New Treatment Available for the Donor Patient

While in some respects RTD programs are similar to oncology research protocols in sharing certain ethical and clinical issues, in other respects they differ markedly. One of the most relevant differences between them concerns the kind of benefit available to participants. RTD takes place after death, which means that no direct benefit, such as new treatments or care, is available for participants. Thus, it becomes very important to analyze the motivations that lead patients to evaluate RTD participation. Since the program takes place after the patient has died, the exclusion of a possible direct benefit to the patient would seem rather trivial to consider. Given this, patients who decide to participate are likely to be animated by a sense of concrete altruism: donating tissues to advance cancer research, so as to contribute to the possibility of developing more effective treatments for the patients of tomorrow. However, although a correct assessment of a risk–benefit ratio for RTD may seem less challenging, the risk of misinterpretation of potential benefits at stake that could impair the entire informed consent process should not be underestimated. As a result, it becomes essential to understand why patients may decide to enter an RTD program.

8.9   Forms of Altruism

It has been previously highlighted how altruism is a central value in the RTD process. However, altruism has several aspects, and it is important to assess whether patients' choices are nourished by a form of altruism that is appropriate in the context. When patients show interest in or willingness to enter an RTD program, it is extremely important to ascertain and understand why they are considering participation. Thus, it becomes crucial to explore the nature of patients' decisions in order to assess the autonomy of their choices and to understand what contributes to building their *substantial* informed consent—that is, what constitutes an informed consent in its first meaning, since the first meaning serves as a benchmark for the requirements of the *formal* meaning of informed consent. In this respect, one issue to focus on is the form of altruism that underpins such a decision. Donating tissues after death for research purposes is a meaningful choice that patients may make, and one that denotes an expression of altruism. Such a choice should reflect a selfless sacrifice, or a gesture made to benefit other individuals, and one without the concrete expectation of a direct benefit in return. In other words, patients considering participation in an RTD program should be motivated by their personal desire to give back to society, to humanity, and to science.

Altruistic motivation applied to cancer patients' decision to enter an RTD program has not been adequately explored in the research literature. Yet this issue is extremely significant because it sheds new light on how altruism influences patients' decision-making. Understanding why people decide to donate would help to improve the recruitment process and to ensure that patients' decisions truly reflect their personal wishes and inclinations. In other words, exploring the role of altruism within the decision to enter an RTD program may provide useful insights into how patients make decisions concerning RTD, which in turn will help to ensure that their choices reflect a set of preference and values.

Altruism in relation to donations has been the subject of intense debate. It might be argued that genuine altruism does not exist, since individuals always maintain an expectation of a benefit at some point and in some way from their altruistic act (Batson 1991). Yet genuine altruism with no expectation of any reciprocation is only one form of altruism. According to the literature, three types of altruism can be identified with regards to biological donations (Quinn et al. 2013). Genuine altruism, which is also termed "pure altruism", occurs when people donate for the sake of good will, with no secondary ends. There are also forms of altruism based on the expectation of a mutual exchange. "Empathy-induced altruism", also known as "group selection theory", describes individuals who decide to donate because they feel part of a group. According to this form of altruism, donation is a way to improve the future welfare of the group of which the donor is part (Batson and Ahmad 2009). A third form of altruism based on mutual exchange is "reciprocal altruism", which envisions elements of expected return for the donors or their relatives (McCullough et al. 2008).

How do these forms of altruism apply to RTD donation? Pure altruism, which is abundantly discussed in Titmuss's work *The Gift Relationship* (1970), involves the selfless desire to help others. Terminal cancer patients may be motivated by genuine altruism and consider RTD as an opportunity to endow various beneficiaries simultaneously by providing enhanced research to their fellow and future cancer sufferers, their descendants, their physicians, nurses, the entire community, and the hospital that provides their care (Quinn et al. 2013). This form of altruism poses no ethical concerns with regards to tissue donation for an RTD program. Since patients are motivated by their genuine desire to perform an act of selfless altruism, there is no need to delve further into their reasons to ensure that their decision to enter an RTD program is not based on the expectation of reciprocation.

The second form of altruism—namely, group selection theory—stems from the wish to further the well-being of the group of which an individual is part. According to group selection theory, altruism is an individual trait that has been cultivated because of the benefits of being part of a group that shares similar characteristics (Caporael 2001; Quinn et al. 2013). This context assumes that an individual will voluntarily act in the best interest of their group, even if the price is personal sacrifice. Previous research has found that an individual diagnosed with an illness or condition may develop a certain feeling of support and solidarity toward others who have the same illness or condition, and hence toward those they believe constitute a group of which they are part (Amir and Haskell 1997; Barr 2006; Quinn et al. 2013; Sanner 2006). In such cases, people feel integrated within a group that shares the same adversities and concerns, and they may express the wish to give back to the group. Cancer patients are an example of this kind of group, and patients who accept their diagnosis and exhibit appropriate coping are more likely to possess a desire to voluntarily give back to the cancer group they feel part of. However, patients who experience difficulty in accepting cancer diagnosis may not exhibit a strong commitment to give back to the community, since they do not feel part of it. Thus, this form of altruism, when supported by a genuine and autonomous desire to give back to the cancer group, does not pose challenging concerns to informed consent for RTD. If the desire to give back to the group reflects the spontaneous wish of a member of the group, this does not raise concerns relating to nudges to autonomy. In other words, that a patient feels the desire to give back to their group by entering an RTD program seems to pose no problem with regard to the autonomy of the patient's decision. Indeed, when patients and families show desire to voluntarily give back to the cancer community, this feeling may be used as a springboard to discuss RTD (Quinn et al. 2013).

The third form of altruism—namely, reciprocal altruism—relies on the idea that there is no selfless altruism: every act is made with an expectation of return. According to the theory of reciprocal altruism, individuals hold an expectation that their beneficial act will be returned in some form to the giver, or to one of their relatives, in the future. Reciprocal altruism thus maintains an underlying principle of mutual obligation. This form of altruism poses concrete ethical concerns if applied within the context of RTD programs. Patients acting with reciprocal altruism may thus decide to enter the program in the expectation that any benefits to others will be reciprocated at some point in the future. In the specific case of an RTD program, reciprocal altruism may take different forms. Patients may expect to be reciprocated by receiving the benefit of a higher level of care or attention at the end of life; or they may decide to enroll because they expect their family to receive, after their death, a higher level of care and treatment as a compensation for the donation.[17] This latter case may be frequent in countries, such as the US, where healthcare is not free but is managed through costly health insurance. In these two hypothetical cases, patients may show interest in entering the program because they expect their act to be reciprocated in some way. However, another possible explanation of why a patient decides to enter the program is that the patient feels obligated to reciprocate and to give back to the hospital or to society for the care received. In fact, while in the first two cases the decision to enter the program is an altruistic act with an expectation of reciprocation, in the third scenario the very act of entering the program represents a way to reciprocate the altruistic act already received by the patient.

The mechanism underpinning every form of reciprocal altruism presents several ethical concerns with regards to RTD programs, since the web of mutual obligations it establishes may prevent patients from making autonomous decisions about informed consent procedures. The three cases considered above present hypothetical situations of conditioned decisions. If patients consent to enter the RTD program because they hope to receive a higher level of care at the end of life, their decision is highly impaired and conditioned by a consideration that should not be taken into account or be a driver to donation. Every patient should be treated and cared for regardless of the decision whether or not to enter an RTD program. Accordingly, if the decision to enter the program is tainted by the chance of a higher quality of care being reserved to the surviving family, the informed consent has been significantly conditioned by aspects that should play no role in such a decision. A patient may decide to consider the option of RTD because they autonomously feel they want to give back to clinical staff or to society by contributing to scientific progress through tissue donation. However, in no circumstances should this choice be imposed, or go against personal preferences and values, by a feeling of duty toward institutions or society arising from the care received as a patient.

8.10   Different Kinds of Benefits

As previously discussed in Chapter 5 above, an extremely influential article by Nancy King (2000) provided a thorough analysis of the different kinds of benefit that may arise from clinical care and clinical research, namely "direct benefit", "collateral benefit", and "aspirational benefit".

Direct benefit arises from receiving an intervention: it is a concrete and effective benefit that is immediately perceived because of the treatment. If I have a cold and I take paracetamol, I will immediately feel relief because of the drug's effect: I am thus experiencing a direct benefit. It is important to focus on this aspect, because an

---

[17]See the next section for a discussion of this issue.

immediate benefit can also be derived from psychological and emotional relief only indirectly caused by the treatment. A patient who enrolls in a Phase I oncology trial may feel immediately relieved because of the feeling that they are fighting cancer instead of giving up. However, such a benefit is not directly associated with the treatment; rather, it is an "indirect" effect of the treatment, which should not be confused with a direct benefit that is directly associated with the treatment. Consequently, King uses the term "collateral benefit" to refer to a benefit arising from being a subject of a study, even if the subject does not get the experimental intervention. This benefit involves, for example, receiving a free physical exam and testing, or free medical care and other extras,[18] or it entails the personal gratification of altruism. In other words, this "indirect" benefit regards every kind of benefit experienced by the patient that arises from the treatment but is not directly related to it. This kind of benefit may be relevant to patients who decide to join an RTD program, because it may provide a personal gratification of altruism.

The third form of benefit is what King calls "aspirational" benefit, which is related to benefits generated by science for future patients. This kind of benefit is the one that best applies to an RTD program. As a result, patients who decide to enter an RTD program should do so concretely because of the benefit their donated tissues will bring to research on new treatments and therapies for future patients. In light of the above analysis, it is worth emphasizing that a misconception of benefits available for the patient may also occur in the framework of RTD (Quinn et al. 2013), even if such a misconception of benefits relating to a postmortem research program might seem bizarre.[19]

An example of misconceptions of benefits relating to RTD is a patient's expectation of being provided with better end-of-life care, better assistance and consideration, or some level of privilege from the clinical staff upon enrollment. Furthermore, some cancer patients may expect their surviving families to be provided with benefits upon their decision to consent to an RTD program, through what I define as "transitive collateral benefit", namely, concrete benefits to relatives in the name of the donation. Such benefits may include better treatments, privileges, or special assistance reserved for the family as a way to reciprocate the consent to tissue retrieval. In a certain sense, this misunderstanding can arise when patients maintain an attitude of reciprocal altruism, because they may expect a benefit when the only benefit concretely provided within the program is either an "aspirational" or a "collateral benefit" associated with the feeling of an altruistic choice. In other words, RTD may provide aspirational benefits—namely, those that benefit future generations of cancer patients—and collateral benefits—namely, positive feelings experienced by donor patients because they decided to donate and believe their pain will be useful to somebody else in the future. Experience of collateral benefit might arise from the feelings of engagement, altruism, and participation that emerge from the agreement

---

[18]These benefits may be particularly appealing in countries like the US where medical assistance depends on health insurance coverage.

[19]I do not venture here into philosophical discussions about certain types of benefit that some believe may occur after death, especially those relating to a belief in an afterlife.

between patients and researchers that RTD consent entails. This feeling of inclusion and of participation might help RTD patients to feel somehow relieved about considering themselves part of a team joining forces against cancer.

Thus, the only benefits that may be associated with RTD are the aspirational benefits to help future patients affected by cancer—this benefit arises after the research takes place—and collateral benefits associated with the feeling of solace and comfort for having contributed to advances in medicine—and these collateral benefits may arise before the research takes place, namely upon consent. Apart from these benefits, patients signing a consent form for RTD programs should not expect benefits in terms of treatments when they are alive, nor should they expect better care, including end-of-life care or some form of privilege for the surviving family.

It is, therefore, fundamental to assess during the informed consent process whether patients considering enrollment in RTD programs clearly understand the program and how it is intended to advance cancer research. Similarly, it is vital that donors are not provided with misperceived benefits or transitive collateral benefits. In this process, the role of the RTD ethicist is crucial.

8.11 Compensation for RTD

The informed consent process should also pay attention to the financial aspect. Let us consider again the comparison with organ donation. Is there an option for donors (or, rather, for their families) to claim some sort of compensation for their donations? Imagine that Mr. X died in a car accident and that his family decided to donate his organs. His heart was assigned to an unnoteworthy artist called Jeff Key. The surgery was successful and saved the artist's life. Imagine further that the artist, a couple of years after the transplant, becomes famous and that one of his creations, *The Flying Cup*, was sold for $58.4 million. Should part of the profit be given to Mr. X's family? It would seem bizarre. The artist could voluntarily offer compensation, but the donor family could certainly make no claim. Now imagine that Mr. Z, an unknown artist in financial difficulties with a wife and three children, has terminal cancer. He decides, supported by his family, to donate cancer tissues with RTD to help future cancer patients. Mr. Z's cancer tissues are inoculated in xenopatients, and, after tests and experiments, they contribute to the development of a new cancer treatment, patented by a pharma company for $7 million. Should any compensation be given by the pharmaceutical company to Mr. Z's family who are in serious financial difficulties? All things considered, if Z and patients like him had not donated their tissues to research, the company would not have patented the same cancer treatment and would not have earned $7 million. This scenario raises ethical concerns with regard to the significant income that retrieved biological material potentially generates for pharmaceutical companies. At the same time, it is also worth noting that pharmaceutical companies invest in scientific research and support expensive trials without having the guarantee that effective treatments will be developed. Nevertheless, for those successful trials that result in patented treatments, the financial return is undoubtedly substantial.

The lack of economic return for donors alongside the huge income for pharmaceutical companies is a significant ethical concern, yet its implications should not impact the altruistic nature of participation in RTD programs. Donation of biological material through RTD within oncology research can appeal to a strong group identity and its nature should be framed within the dimension of an altruistic action from today's cancer patients to those of tomorrow. In this framework, considering the long (and uncertain) research process, even if, hypothetically, compensation was extended to RTD donors, it would be difficult to establish the original ownership of the single samples used over years of research and to precisely determine the extent to which samples were decisive for the development of the patented treatment.

In light of this, the chance offered to RTD patients to choose which groups of research their tissues might be used for can be helpful. Moreover, the imbalance between donors and pharmaceutical companies could be mitigated through alternative means, such as companies being invited to invest part of the income resulting from patented treatments in social programs that benefit the families of cancer patients and, indirectly, the patients themselves (Buyx and Prainsack 2018; Prainsack and Buyx 2018). However, although the ethical implications of such an imbalance are recognized, analysis of it goes beyond the scope of the present work.

RTD donation should reflect an altruistic donation by cancer patients that aims to contribute to scientific advances for future patients; thus, no financial compensation should be envisioned. Accordingly, patients should be made aware that no remuneration is provided for participation in RTD programs; nor can patients expect their families to receive any share of future profits that arise from the tissue retrieval. At the same time, postmortem research should not impose extra costs on the subject's estate or on their relatives. Given its novelty and potentially controversial aspects, participation in RTD programs should be recognized as motivated by altruism. The donation envisioned by RTD programs presents an altruistic opportunity: it should remain an opportunity for self-motivated donation without any additional economic implications for the deceased patient. Financial aid may solely involve body transportation after tissue retrieval in order for the burial ceremony to take place, as the next section addresses.

8.12 Timing and Logistics of Cadaver's Return to Families

One of the greatest disadvantages of RTD involves timing, because there is a narrow window between the patient's death and tissue retrieval. First, it is important to consider where the patient dies. If the patient dies in the facility where the RTD program takes place, the logistics are straightforward. However, a patient may be at home or in a hospice (Howell and Lutz 2008), with transportation needing to be arranged in advance and having to take place in a narrow window of time. Consequently, the research protocol associated with RTD raises several ethical concerns in relation to time. The time needed for tissue retrieval is time in which the body is kept away from relatives. This may dramatically exacerbate suffering and create complications for the organization of the funeral. Thus, the RTD team is strongly advised to keep the retrieval process as limited as possible to return the body to families and let them proceed with funeral arrangements. Minimizing the time it takes for tissue retrieval means that families can take care of the logistical issues and be able to mourn for their deceased loved one as soon as possible. Therefore, a definitive schedule that includes timing of the body's return should be promptly communicated and made available to families in order for them to organize the funeral. Family members should be updated about the proceeding of the operations and, although any delay in the return of the body should be avoided, family members must be warned promptly if such an unpleasant event should occur.

Given these issues, the RTD ethicist should provide the family with accurate and punctual information, tailored to the audience, about the procedures of tissue retrieval. RTD ethicists should be the only figures in charge of reporting and communication with families in this phase: in such an extremely delicate moment for relatives, any overlapping of information—with the possibility of conflicting information coming from different people not in charge of communication and not adequately trained in it—can lead to misunderstandings and misinterpretations by the grieving relatives, and hence to an unnecessary increase in their stress and anxiety. Such circumstances should be absolutely avoided.

As a communication hub, the RTD ethicist can help to reassure relatives in a difficult moment in their lives. In particular, reassurance can be provided to families that the body of their loved one, from which they were separated, is not abandoned in some aisle or does not lie in an empty laboratory but is being cared for and will be returned to the family as soon as possible. Especially at this time, the healthcare professionals' team must respect their roles, offering support to relatives if needed; hence, they should work to ensure that the body is returned to family members in accordance with the scheduled time.

Relatives should be made aware that any research costs for body preparation and transportation (to the facility where tissue collection happens and from this facility to the funeral home, crematorium, or other place requested by the family) should not be charged to families or to the deceased's estate; rather, logistics should be managed by the research center responsible for the retrieved tissues. Patients and families should be made aware that "replenishment" of the body after tissue retrieval is granted with full coverage by the program.

The centrality of the RTD ethicist in this phase is emblematic of the importance this individual has within the clinical context and, in particular, within RTD. Commencing with the support provided to the donor, and encompassing the decision-making process, the resolution of conflicts by using cautious and respectful mediation, and the cherishing of family members in the collection of the tissues, the role of the clinical ethicist truly stands at the heart of the RTD program.

Although the clinical ethicist provides support to the health team, family members, and the patient, it is to the latter that they must be most dedicated. Indeed, the main goal of the RTD ethicist is to ensure that the patient is in the condition to make a choice on tissue donation that reflects the patient's own preferences and values, regardless of external influences that may occur, by ensuring that the informed consent process is valid not only in a *formal* sense but also in a *substantial* one. Given these premises, after building a legally recognized path that gives cancer patients the chance to donate their tissues to research—thus an informed consent in its *formal* sense—it is precisely the RTD ethicist who ensures that this informed consent process is also valid in a *substantial* sense.

References

Amir, M., and E. Haskell. 1997. Organ donation: Who is not willing to commit? Psychological factors influencing the individual's decision to commit to organ donation after death. *The International Journal of Behavioral Medicine* 4 (3): 2015–2229.

Andorno, R. 2004. The right not to know: An autonomy based approach. *Journal of Medical Ethics* 30 (5): 435–439.

Barr, M. 2006. I'm not really read up on genetics: Biobanks and the social context of informed consent. *BioSocieties* 1 (2): 251–262.

Batson, C.D. 1991. *The altruism question: Toward a social-psychological answer*. New York: Erlbaum.

Batson, C.D., and N.Y. Ahmad. 2009. Empathy induced altruism: A threat to the collective good. In *Altruism and prosocial behaviors in groups: Advances in group processes*, ed. E.J. Lawler, 1–23. Bingley: Emerald Group Publishing.

BBMRI network. www.bbmri-eric.eu.

Berkman, B.E., and S.C. Hull. 2014. The "right not to know" in the genomic era: Time to break from tradition? *The American Journal of Bioethics* 14 (3): 28–31.

Bookman, E.B., et al. 2006. Reporting genetic results in research studies: Summary and recommendations of an NHLBI working group. *American Journal of Medical Genetics* Part A 140 (10): 1033–1040.

Boyle, T., et al. 2020. A community based lung cancer rapid tissue donation protocol provides high quality drug-resistant specimens for proteogenomic analyses. *Cancer Medicine* 9 (1): 225–237.

Buyx, A., and B. Prainsack. 2018. Solidarity can make a difference: addressing transformations in healthcare, demographics and technological replacement. *Bioethics* 32 (9): 537–540.

Cadigan, R.J., et al. 2011. The meaning of genetic research results: Reflections from individuals with and without a known genetic disorder. *The Journal of Empirical Research on Human Research Ethics* 6 (4): 30–40.

Caporael, L.R. 2001. Evolutionary psychology: Toward a unifying theory and a hybrid science. *Annual Review of Psychology* 52 (1): 607–628.

Chadwick, R.F. 2004. The right not to know: A challenge for accurate self-assessment. *Philosophy, Psychiatry, & Psychology* 11 (4): 299–301.

Clift, K.E., et al. 2015. Patients' views on incidental findings from clinical exome sequencing. *Applied and Translational Genomics* 4: 38–43.

Commissione per l'Etica della Ricerca e la Bioetica del CNR. 2018. Incidental findings nella ricerca scientifica. Indicazioni e criteri per le scienze e tecnologie -omiche. www.cnr.it/it/ethics.

Croyle, R.T., and C. Lerman. 1999. Risk communication in genetic testing for cancer susceptibility. *JNCI Monographs* 25: 59–66.

European Commission, Data protection in the EU. 2016. Regulation (EU) 2016/679 of the European parliament and of the Council of 27 April 2016 on the Protection of Natural Persons with Regard to the Processing of Personal Data and on the Free Movement of Such Data, and Repealing Directive 95/46/EC (General Data Protection Regulation). https://eur-lex.europa.eu/eli/reg/2016/679/oj.

European Commission, Justice and Consumers. 2016. Article 29 data protection working party "Guidelines on Transparency under Regulation 2016/679". https://ec.europa.eu/newsroom/articl e29/item-detail.cfm?item_id=622227.

Gray, S.W., et al. 2016. Oncologists' and cancer patients' views on whole-exome sequencing and incidental findings: Results from the CanSeq Study. *Genetics in Medicine* 18 (10): 1011–1019.

Green, R.C., et al. 2013. ACMG recommendations for reporting of incidental findings in clinical exome and genome sequencing. *Genetics in Medicine* 15 (7): 565–574.

Hall, A., et al. 2013. *Managing incidental and pertinent findings from WGS in the 100,000 genome project.* PHG Foundation. https://www.phgfoundation.org/documents/326_1369298828.pdf.

Harris, J., and K. Keywood. 2001. Ignorance, information and autonomy. *Theoretical Medicine and Bioethics* 22 (5): 415–436.

Howell, D.D., and S. Lutz. 2008. Hospice referral: An important responsibility of the oncologist. *Journal of Oncology Practice* 4 (6): 303–304.

Jelsig, A.M., et al. 2015. Research participants in NGS studies want to know about incidental findings. *European Journal of Human Genetics* 23 (10): 1423–1426.

Kahneman, D., and A. Tversky. 1979. Prospect theory: An analysis of decision under risk. *Econometrica* 47 (2): 263–292.

Kalia, S.S., et al. 2017. Recommendations for reporting of secondary findings in clinical exome and genome sequencing, 2016 Update (ACMG SF v2.0): A policy statement of the American College of Medical Genetics and Genomics. *Genetics in Medicine* 19 (2): 249–255.

Kaphingst, K.A., et al. 2016. Preferences for return of incidental findings from genome sequencing among women diagnosed with breast cancer at a young age: Young breast cancer patients' preferences for return of results. *Clinical Genetics* 89 (3): 378–384.

Kaye, J., et al. 2015. Dynamic consent: A patient interface for twenty-first century research networks. *European Journal of Human Genetics* 23: 141–146.

King, N.M. 2000. Defining and describing benefit appropriately in clinical trials. *The Journal of Law, Medicine and Ethics* 28 (4): 332–343.

Kirschen, M.P., et al. 2006. Subjects' expectations in neuroimaging research. *Journal of Magnetic Resonance Imaging* 23 (2): 205–209.

Kleiderman, E., et al. 2013. Returning incidental findings from genetic research to children: Views of parents of children affected by rare diseases. *Journal of Medical Ethics* 40 (10): 691–696.

Laurie, G.T. 1999. In defence of ignorance: Genetic information and the right not to know. *European Journal of Health Law* 6: 119–132.

Lindell, K.O., et al. 2006. Lessons from our patients: Development of a warm autopsy program. *PLoS Medicine* 3 (7): e234.

Mandava, A., et al. 2015. When should genome researchers disclose misattributed parentage? *Hastings Center Report* 45 (4): 28–36.

McCullough, M.E., et al. 2008. An adaptation for altruism: The social causes, social effects, and social evolution of gratitude. *Current Directions in Psychological Science* 17 (4): 281–285.

McIntyre, J., et al. 2013. Stakeholder perceptions of thoracic rapid tissue donation: An exploratory study. *Social Science and Medicine* 99: 35–41.

Meacham, M.C., et al. 2010. Researcher perspectives on disclosure of incidental findings in genetic research. *The Journal of Empirical Research on Human Research Ethics* 5 (3): 31–41.

Murphy, J., et al. 2008. Public expectations for return of results from large-cohort genetic research. *The American Journal of Bioethics* 8 (11): 36–43.

Ost, D.E. 1984. The "right" not to know. *Journal of Medicine and Philosophy* 9 (3): 301–312.

Prainsack, B., and A. Buyx. 2018. The value of work: Addressing the future of work through the lens of solidarity. *Bioethics* 32 (9): 585–592.

Presidential Commission for the Study of Bioethical Issues. 2013. Anticipate and communicate: Ethical management of incidental and secondary findings in the clinical, research, and direct-to-consumer contexts. https://www.genome.gov/Pages/PolicyEthics/HealthIssues/Anticipate_C ommunicate.pdf.

Quinn, G.P., et al. 2013. Altruism in terminal cancer patients and rapid tissue donation program: Does the theory apply? *Medicine, Health Care and Philosophy* 16 (4): 857–864.

Ravitsky, V., and B.S. Wilfond. 2006. Disclosing individual genetic results to research participants. *The American Journal of Bioethics* 6 (6): 8–17.

Rhodes, R., and K.L. Capitulo. 2006. Genetic testing: is there a right not to know? *MCN; American Journal of Maternal Child Nursing* 31 (3): 145.

Robertson, S., and J. Savelescu. 2001. Is there a case in favour of predictive genetic testing in young children? *Bioethics* 15 (1): 26–49.

Royal College of Radiologists. 2011. Management of incidental findings detected during research imaging. https://www.rcr.ac.uk/system/files/publication/field_publication_files/BFCR%2811% 298_ethics.pdf.

Sanner, M. 2006. People's attitudes and reactions to organ donation. *Mortality* 11 (2): 133–150.

Schaefer, G.O., and J. Savulescu. 2018. The right to know: A revised standard for reporting incidental findings. *Hastings Center Report* 48 (2): 22–32.

Shalowitz, D.I., and F.D. Miller. 2005. Disclosing individual results of clinical research: Implications of respect for participants. *JAMA* 294 (6): 737–740.

Shaw, M.W. 1987. Testing for the Huntington gene: A right to know, a right not to know, or a duty to know. *American Journal of Medical Genetics* 26 (2): 243–246.

Titmuss, R.M. 1970. *The gift relationship: From human blood to social policy.* London: London School of Economics Books.

Townsend, A., et al. 2012. I want to know what's in Pandora's box: Comparing stake-holder perspectives on incidental findings in clinical whole genomic sequencing. *American Journal of Medical Genetics* 158A (10): 2519–2525.

UNESCO. 1997. Declaration of the human genome and human rights, adopted by the general conference of the United Nations Educational, Scientific and Cultural Organization at its 29th Session on 11 November 1997. https://en.unesco.org/themes/ethics-science-and-technology/human-gen ome-and-human-rights.

Viberg, J., et al. 2016. Freedom of choice about incidental findings can frustrate participants' true preferences. *Bioethics* 30 (3): 203–209.

Wilkinson, T.M. 2005. Individual and family consent to organ and tissue donation: Is the current position coherent? *Journal of Medical Ethics* 31 (10): 587–590.

Wilson, J. 2005. To know or not to know? Genetic ignorance, autonomy and paternalism. *Bioethics* 19 (5–6): 492–504.

Wolf, S.M., et al. 2008. Managing incidental findings in human subjects research: Analysis and recommendations. *The Journal of Law, Medicine and Ethics* 36 (2): 219–248.

Conclusion

In a world where cancer is the second leading cause of death, early diagnosis and effective treatments are fundamental to improving the survival rates. Rapid advances in medical research, particularly that which focuses on "omics" technologies, are opening up new treatment opportunities and cutting-edge prevention approaches. Precision medicine is an especially promising approach to disease prevention and treatment; in this approach, individual variability in genes, environment, and lifestyle plays a key role. When applied to oncology, precision medicine aims to tailor medical treatment to the individual characteristics of each patient. To do this, it relies on a deep understanding of the molecular landscape of tumors. However, in order to advance this understanding, it is necessary to have access to high-quality cancer tissues. Obtaining a sufficient quantity of quality tissue using traditional biopsies in living patients is unfeasible; postmortem retrieval of tissues, on the other hand, offers significantly better opportunities for collection, and hence for oncological research. Rapid Tissue Donation (RTD) is a procedure that offers just such opportunities, since it involves the procurement of "fresh" tissue within 2–6 h of the death of a cancer patient.

However, as this book has discussed, RTD raises unprecedented issues relating to the informed consent process. These issues have been considered by using a dual-sense framework to interpret informed consent. These two senses of informed consent—its formal sense and its substantial sense—constitute the thread that has run through this book.

I have argued that an informed consent to regulate RTD is necessary. In doing so, I have challenged arguments according to which the dead can be viewed as *res nullius*, as abandoned cadavers, or as natural resources to which every (living) moral agent has equal entitlement. On the contrary, the dead are bodies of individuals who have led a life, whether long or short, that was shaped by choices, values, and aspirations. This former aliveness—together with all the characteristics, experiences, and qualities of life that go with that—is an important feature of the dead, and one that is overlooked in many of the scholarly arguments. Consequently, I have referred to the dead as the

C. Mannelli, *The Ethics of Rapid Tissue Donation (RTD)*, The International Library of Bioethics 85, https://doi.org/10.1007/978-3-030-67201-0

"once-alive" to express that former aliveness and to bridge it to their condition of death. By viewing the dead as the once-alive—by connecting them to their former aliveness—it becomes important to honor the wishes they expressed during their lives.

Is it, however, always necessary to honor the wishes of the once-alive? I have argued that it is not. Their wishes and preferences should be honored only when they do not jeopardize other living individuals. Thus, the expressed wishes (if any) of the once-alive concerning the donation of cancer tissues to oncology research should be honored *with a condition*. In relation to this, I have considered the view that a refusal to donate jeopardizes living individuals, since it might be claimed that it denies to researchers the possibility of developing new treatments. In other words, some might argue that refusal of consent fails the condition outlined above of not jeopardizing the living. My analysis shows that this argument is unjustified. Once an RTD program has been established and adequately implemented, cancer patients who decide to enroll will provide sufficient material for this branch of medicine to develop. Hence, those who refuse to donate tissues through RTD would not jeopardize the living. My position has been developed by comparing the specific framework of RTD with that of organ donation. Whereas the latter provides society with immediate, certain, and concrete benefits, the benefits to society of RTD, although highly promising, are neither immediate nor certain. Given the abstractness of the benefits directly associated with RTD, an argument for conscription of donation, which would lead to a lack of informed consent governing this procedure, is not warranted.

The need for an informed consent to regulate RTD inevitably leads to questions about how informed consent is structured and implemented in the real world of the clinical setting, and particularly how to do so in a way that complies with both the substantial and the formal senses of informed consent. I have approached these issues by considering ethical issues relating to information, comprehension, and voluntariness, all of which are relevant to general oncology research, and which provide a basis for analyzing the issues raised by RTD. My discussion has brought out the importance of various considerations—such as the right timing for a consultation, and the choice of spokesperson—in order to ensure that prospective participants in an RTD program are provided with the information they need to make an informed choice in both the *formal* and *substantial* senses of the term. The informed consent procedure for RTD that I have developed in this book has a specific focus on the most delicate aspects that this procedure can bring up. The envisioned procedure does not take into account a particular legal framework. Rather, the book's analysis intentionally leans toward creating a general framework, based on ethical and philosophical analysis, rather than to setting out policy. By delineating and analyzing within this framework the ethically relevant issues raised by RTD, it should then be possible to adapt and contextualize them according to a specific legal framework.

An important part of my argument has been that a dedicated RTD ethicist is a key figure in the informed consent process. The aim of the RTD ethicist is to support patients, families, and healthcare professionals in the decision-making process, in mediation, and in conflict resolution. Given the difficult, sensitive, and delicate issues associated with RTD, the RTD ethicist has a crucial role in ensuring the informed consent process works effectively and, as far as possible, honors the wishes of patients, while also supporting families and healthcare professionals through the often difficult decision-making entailed by RTD.

I have, therefore, proposed an informed consent in the *formal* sense of its meaning, which constitutes the outline of a legally recognized path for cancer patients who wish to donate their tissues to research and who want to formalize their wishes. At the same time, my definition of the requirement of an informed consent in its *formal* sense has been predicated on keeping the autonomous authorization of the patient at the forefront; in other words—and in line with my overall argument about the need to consider both senses of informed consent as equally important—my outline of an informed consent in its *formal* sense is also based on a consideration of informed consent in its *substantial* sense.

The ultimate aim of this work has been to analyze the unprecedented ethical issues posed by this medical technique. The RTD framework is just one of many emerging areas of medicine that would benefit from an ethical analysis. The issues raised by RTD, as by other novel medical techniques and procedures, are not limited to the healthcare context, since they also have wider ethical, social, and psychological implications, to name just a few areas. These are important issues to explore, not least because, as in the case of RTD, we might entertain the ambitious hope that the value of an ethical approach, and its vital contribution to advances in research and healthcare, will be increasingly acknowledged and implemented for patients, families, healthcare professionals, and society as a whole.

Appendices[1]

---

[1]This section contains examples of informed consent forms and information sheets dedicated to RTD. This is a possible solution (but by no means the only feasible, or even the best one) to govern the informed consent process within the particular framework of RTD.

1.RTD BROCHURE

Rapid Tissue Donation - RTD

WHAT IS RTD?

Rapid Tissue Donation (RTD) is the collection of tissues shortly after (generally between 2 and 6 hours) the death of a cancer patient. This technique offers researchers an extraordinary opportunity to study cancer onset and development in order to translate these discoveries into possible new treatments to fight cancer.

WHO CAN BECOME A DONOR?

Any cancer patient can become a donor.

HOW IS RTD PERFORMED?

RTD involves the retrieval of small portions of cancer tissues from both the primary tumor and the metastases after the death of a patient.

HOW DO I CONSENT?

Your treating physician will explain the procedure, and then, if you are interested in donation, a clinical ethicist will address all your questions.

IS THERE A COMPENSATION?

No, there is no compensation for donation. This procedure will incur no financial cost to the donor or the family.

WHAT ARE THE RISKS?

There are no risks as the retrieval of tissues happens after a patient's death.

WHAT ARE THE BENEFITS?

There are no direct benefits for participating patients or their families. However, tissue donation may help scientists find treatments for future patients.

FOR FURTHER INFORMATION ON RTD ASK YOUR TREATING PHYSICIAN, CONTACT THE PROGRAM AT RTD@GMAIL.COM OR VISIT WWW.RTD.EU

2. RTD IC2 – Competent Adults Information Sheet & Informed Consent Form

RTD PROGRAM – INFORMATION SHEET

Dear Sir/Madam,

You are reading this sheet as you requested information on Rapid Tissue Donation (RTD). This sheet contains detailed and relevant information.

Before you make any decision, take your time to read and fully understand this document. This content will be explained to you by the RTD ethicist who will devote the appropriate time to talk to you and to respond to your questions. There are no insignificant questions; any doubt is worth discussing with the RTD ethicist.

What is Rapid Tissue Donation?

Rapid Tissue Donation (RTD) is a program of tissue donation specifically for cancer patients.

RTD involves the retrieval of cancer tissues and takes place after death. RTD offers researchers the chance to study the characteristics of tumor tissues to understand the development of tumors and to advance research. To this end, it is crucial for tissues to be collected straight after death so that they do not lose their specific characteristics. The window of time required for collection is fixed at two to six hours following death.

Is my participation in RTD Mandatory?

No, your participation is not mandatory. It is your decision. You can discuss this option with your family, your friends, and your physician. If you have any questions, please ask the RTD ethicist, whose contact details are provided at the end of this sheet.

What risks are involved for me?

There are no risks for you because tissue collection takes place after death. The retrieval of tissues is performed by a physician who specializes in pathology and is assisted by the health care staff. The procedure follows rigorous medical and ethical protocols.

What benefits are involved?

There are no direct benefits for you as tissue collection takes place after death. There will be no benefits for your family or your relatives now. However, your tissues will contribute to the development of cancer research and might benefit future patients like you.

What happens if I consent?

If you consent to the program by signing this form, you will be asked to confirm or revoke your consent later. You are free to change your mind whenever you want by contacting the RTD ethicist, whose contact details are provided at the end of this form. Your treatment and care will not be impacted by your choice. You will be supported with the best available treatments and care regardless of your choice.

What happens if I refuse?

If you refuse to enter the program, you do not have to sign this form. Nobody will ask again. If you change your mind, and you want further information on RTD, you can put any questions to the RTD ethicist, whose contact details are provided at the end of this sheet. If you refuse to enter the program, your treatment will not be impacted in any way by your choice. You will be supported with the best available treatment and care regardless of your choice, and your refusal will in no way be interpreted as a lack of trust.

Can I change my mind?

Yes. You can change your mind whenever you want before tissues are collected. It is your right. Just tell the RTD ethicist whose contact details are provided at the end of this sheet.

Please note that consent withdrawal is not retroactive. So, if your next of kin revokes consent after the tissues have been collected and perhaps used in some research, the results of the experiments will not be discarded. However, after consent withdrawal, any further use of the tissues and related data will be prohibited.

How will my tissues be used?

Collected tissues will be used for research aims in order to advance scientific development and translate it into better treatments for patients. Research is a complex and rigorous process that requires time and patience. This means that tissues collected today might reveal themselves useful tomorrow for aims that were not imaginable today. For this reason, it is difficult to tell you the precise use for your tissues.

However, we understand that it is important for you to know how your tissues will be used. You can choose the aims for which your tissues will be used by checking the corresponding box(es).

Who is responsible for the use of my tissues?

The RTD program is responsible for the retrieved tissues, which will be managed responsibly and in compliance with ethical and legal frameworks. The RTD program will respect your choices.

Will my tissues be commercialized?

Your tissues cannot be sold for profit. However, your tissues might be used for research that will lead to commercial products, such as diagnostic kits if you have checked the corresponding box. Any financial income resulting from these activities will not constitute grounds for a claim by you or your family.

How long will my tissues be stored?

At the present time, it is not possible to quantify a precise storage period, as tissues will be kept until their quality is suitable to obtain reliable results.

How are tissues conserved?

In order to safeguard your privacy, retrieved tissues will be pseudonymized, which means that they will be associated with a code. Only the RTD program will know which code corresponds with which name.

How will my decision affect my relatives?
If you are thinking about entering the RTD program, it is advisable that you share your thoughts and your decision with your relatives, so that they can be prepared. This is also important because tissues contain genetic information, and research on your tissues might be of interest to your relatives if incidental findings occur.

What are incidental findings?
Incidental findings are observations on your tissues that may occur during research but that are not related to the goals of research. These observations may reveal susceptibility to developing a particular disease that may be relevant to your blood relatives. They can decide to be informed of any prospective incidental findings or not.

**We thank you for the attention and time you have devoted to reading this document.
If you decide to participate in the study, please complete and sign the form below. A copy of the information sheet will be given to you.**

CONTACTS:
RTD PROGRAM GENERAL INFORMATION   dr. Name, Surname, mobile, email
RTD ETHICIST   dr. Name, Surname, mobile, email

RTD IC2 – Competent Adults Information Sheet & Informed Consent Form

♨️RTD

RTD – INFORMED CONSENT FORM

I, the undersigned .... born in............... on...............

Hereby declare to have:

read and understood the information sheet whose contents have been explained by the RTD ethicist Dr.
...............................;
had time to discuss my decision with other persons;

had time to ask questions and received clarification about the program.

Therefore, I voluntarily

O   consent   O   refuse

to participate in the RTD program. To this end, I:

- consent to the use of my tissues for:

*Check the box next to the aim(s) **you want your tissues to be used for** (you can check more than one box)*

O   **oncology research only**

O   **research for all diseases**

O   **commercial use of the sample (diagnostic kits, etc.)**

O   **non-medical areas of research (cosmetics, etc.)**

O   **future uses now unknown**

....Participant signature
....Signature of the RTD ethicist collecting consent

Next of kin

.... Next of kin Name, Surname

.... Next of kin Signature

Date.................... Place...............

RTD PROGRAM GENERAL INFORMATION   dr. Name, Surname, mobile, email

RTD ETHICIST   dr. Name, Surname, mobile, email

3. RTD IC2_LR – Adults with limited decision making capacity- Information Sheet & Informed Consent Form

RTD

RTD PROGRAM – INFORMATION SHEET

Dear Sir/Madam,

You are reading this sheet as you requested information on Rapid Tissue Donation (RTD). This sheet contains detailed and relevant information.
Before you make any decision for your ward, take your time to read and fully understand this document. This content will be explained to you by the RTD ethicist who will devote the appropriate time to talk to you and to respond to your questions. There are no insignificant questions; any doubt is worth discussing with the RTD ethicist.

What is Rapid Tissue Donation?

Rapid Tissue Donation (RTD) is a program of tissue donation specifically for cancer patients.
RTD involves the retrieval of cancer tissues and takes place after death. RTD offers researchers the chance to study the characteristics of tumor tissues to understand the development of tumors and to advance research. To this end, it is crucial for tissues to be collected straight after death so that they do not lose their specific characteristics. The window of time required for collection is fixed at two to six hours following death of the patient.

Is the participation of your ward in RTD Mandatory?

No, the participation of your ward is not mandatory. You can discuss this option with your ward, the family and relatives of your ward, or with other physicians. If you have any questions, please ask the RTD ethicist, whose contact details are provided at the end of this sheet.

What risks are involved for my ward?

There are no risks for your ward as tissue collection takes place after death. The retrieval of tissues is performed by a physician who specializes in pathology and is assisted by the health care staff. The procedure follows rigorous medical and ethical protocols.

What benefits are involved?

There are no direct benefits for your ward as tissue collection takes place after death. There will be no benefits for the family or relatives of your ward. However, their tissues will contribute to the development of cancer research and might benefit future patients.

What happens if I consent?

If you consent to the program by signing this form, you will be asked to confirm or revoke your consent later. You are free to change your mind whenever you want by contacting the RTD ethicist, whose contact details are provided at the end of this form. The treatment and care of your ward will not be impacted in any way by your choice. Your ward will be supported with the best available treatment regardless of the choice made.

What happens if I refuse?
If you refuse to enter this program, you do not have to sign this form. Nobody will ask again. If you change your mind, and you want further information on RTD, you can put any questions to the RTD ethicist, whose contact details are provided at the end of this sheet. If you refuse to enter the program, the treatment of your ward will not be impacted by the choice made. Your ward will be cared for with the best available treatments regardless of your choice, and your refusal will in no way be interpreted as a lack of trust.

Can I change my mind?
Yes. You can change your mind whenever you want before tissues are collected. It is your right. Just tell the RTD ethicist whose contact details are provided at the end of this sheet.

Please note that consent withdrawal is not retroactive. So, if you change your mind after the tissues have been collected and perhaps used in some research, the results of the experiments will not be discarded. However, after you have withdrawn consent, any further use of the tissues and related data will be prohibited.

How will the tissues of my ward be used?
Collected tissues will be used for research aims in order to advance scientific development and translate it into better treatments for patients. Research is a complex and rigorous process that requires time and patience. This means that tissues collected today might reveal themselves useful tomorrow for aims that were not imaginable today. For this reason, it is difficult to tell you the precise use for your ward's tissues. However, we understand that it is important for you to know how the tissues will be used. You can choose the aims for which the tissues of your ward will be used by checking the corresponding box(es).

Who is responsible for the use of retrieved tissues?
The RTD program is responsible for the retrieved tissues, which will be managed with responsibility and in compliance with ethical and legal frameworks. The RTD program will respect your choices.

Will the tissues of my ward be commercialized?
Retrieved tissues cannot be sold for profit. However, tissues might be used for research that will lead to commercial products, such as diagnostic kits if you have checked the corresponding box. Any financial income resulting from these activities will not constitute grounds for a claim by your ward or their family.

How long will retrieved tissues be stored?
At the present time, it is not possible to quantify a precise storage period, as tissues will be kept until their quality is suitable to obtain reliable results.

How are tissues conserved?
In order to safeguard the privacy of your ward, retrieved tissues will be pseudonymized, which means that they will be associated with a code. Only the RTD program will know which code corresponds with which name.

How will my decision affect my ward's relatives?
If you are thinking about consenting to the RTD program, you might consider sharing your thoughts with the family of your ward so that they can be prepared. This is also important because tissues contain genetic information, and research on your ward's tissues might be of interest to relatives if incidental findings occur.

What are incidental findings?

Incidental findings are observations on retrieved tissues that may occur during research but that are not related to the goals of research. These observations may reveal susceptibility to developing a particular disease that may be relevant to the blood relatives of your ward. Blood relatives can decide to be informed of any prospective incidental findings or not.

We thank you for the attention and time you have devoted to reading this document.
If you decide to have your ward participate in RTD, please complete and sign the form below.
A copy of the information sheet will be given to you.

CONTACTS:
RTD PROGRAM GENERAL INFORMATION   dr. Name, Surname, mobile, email
RTD ETHICIST   dr. Name, Surname, mobile, email

RTD IC2_LR – Adults with limited decision making capacity- Information Sheet & Informed Consent Form

❧RTD

RTD PROGRAM – INFORMED CONSENT FORM

I, the undersigned ....…. born in…............. on…............

Legal representative of:
name and surname......... born in .... on..........

Hereby declare to have:

read and understood the information sheet whose contents have been explained by the RTD ethicist
Dr. …...........................;
had time to discuss my decision with other persons;

had time to ask questions and received clarification about the program.

Therefore, I voluntarily

O   consent   O   refuse

To have my ward participate in the RTD program. To this end, I:

- consent to the use of my ward's tissues for:

Check the box next to the aim(s) you want your ward's tissues to be used for (you can check more than one box)

O   **oncology research only**

O   **research for all diseases**

O   **commercial use of the sample (diagnostic kits, etc.)**

O   **non-medical areas of research (cosmetics, etc.)**

O   **future uses now unknown**

...................................Legal representative signature

.....................................Signature of the RTD ethicist collecting consent

Date.................. Place..............

RTD PROGRAM GENERAL INFORMATION   dr. Name, Surname, mobile, email
RTD ETHICIST   dr. Name, Surname, mobile, email

4. RTD IC2_PG – Parents/Guardians Information Sheet & Informed Consent Form

❧RTD

INFORMATION SHEET– RTD PROGRAM

Dear Parents/Guardians,

You are reading this sheet as you requested information on Rapid Tissue Donation (RTD). This sheet contains detailed and relevant information.

Before you make any decision for your child, take your time to read and fully understand this paper. This content will be explained to you by the RTD ethicist who will devote the appropriate time to talk to you and respond to your questions. There are no insignificant questions; all questions are worth asking.

What is Rapid Tissue Donation?

Rapid Tissue Donation (RTD) is a program of tissue donation specifically for cancer patients.

RTD involves the retrieval of cancer tissues and takes place after death. RTD offers researchers the chance to study the characteristics of tumor tissues to understand the development of tumors and to advance research. To this end, it is crucial for tissues to be collected straight after death so that they do not lose their specific characteristics. The window of time required for collection is fixed at two to six hours following the death of the patient.

Is the participation of my child in RTD mandatory?

No, the participation of your child is not mandatory. It is your decision. You can discuss this option with your child, with other members of the family, and with other physicians. If you have any questions, please ask the RTD ethicist, whose contact details are provided at the end of this sheet.

What risks are involved for my child?

There are no risks for your child as tissue collection takes place after death. The retrieval of tissues is performed by a physician who specializes in pathology and is assisted by the health care staff. The procedure follows rigorous medical and ethical protocols.

What benefits are involved?

There are no direct benefits for your child as tissue collection takes place after death. There will be no benefits for the child's family or relatives. However, their tissues will contribute to the development of cancer research and might benefit future patients.

What happens if I consent?

If you consent to the program by signing this form, you will be asked to confirm or revoke your consent later. You are free to change your mind whenever you want by contacting the RTD ethicist, whose contact details are provided at the end of this form. The treatment and care of your child will not be affected in any way by your choice. Your child will be cared for with the best available treatment regardless of your choice.

What happens if I refuse?

If you refuse to enter this program, you do not have to sign this form. Nobody will ask again. If you change your mind, and you want further information on RTD, you can put any questions to the RTD ethicist, whose contact details are provided at the end of this form. If you refuse to enter the program, the care and treatment of your child will not be affected in any way by your choice. Your child will be cared for with the best available treatment regardless of your choice, and your refusal will in no way be interpreted as a lack of trust.

Can I change my mind?

Yes. You can change your mind whenever you want before tissues are collected. It is your right. Just tell the RTD ethicist whose contact details are provided at the end of this sheet.

Please note that consent withdrawal is not retroactive. So, if you change your mind after the tissues have been collected and perhaps used in some research, the results of the experiments will not be discarded. However, after you have withdrawn consent, any further use of the tissues and related data will be prohibited.

How will the tissues of my child be used?

Collected tissues will be used for research aims in order to advance scientific development and translate it into better treatments for patients. Research is a complex and rigorous process that requires time and patience. This means that tissues collected today might reveal themselves useful tomorrow for aims that are not imaginable today. For this reason, it is difficult to tell you precisely what your child's tissues will be used for.

However, we understand that it is important for you to know how the tissues will be used. You can choose the aims for which your child's tissues will be used by checking the corresponding box(es).

Who is responsible for the use of retrieved tissues?

The RTD program is responsible for the retrieved tissues, which will be managed responsibly and in compliance with ethical and legal frameworks. The RTD program will respect your choices.

Will my child's tissues be commercialized?

Retrieved tissues cannot be sold for profit. However, tissues might be used for research that will lead to commercial products, such as diagnostic kits if you have checked the corresponding box. Neither your child nor their family will be able to make any financial claim on income resulting from these activities.

How long will retrieved tissues be stored?

At the present time, it is not possible to quantify a precise storage period as tissues will be kept until their quality is suitable to obtain reliable results.

How are tissues conserved?

In order to safeguard the privacy of your child, retrieved tissues will be pseudonymized, which means that they will be associated with a unique code. Only the RTD program knows which code corresponds to which name.

How will my decision affect us and my child's relatives?

If you are thinking about consenting to the RTD program, you might consider sharing your thoughts you're your child's other family members so that they can be prepared. This is also important because retrieved tissues contain genetic information, and research on your child's tissues might be of interest to relatives if incidental findings occur.

What are incidental findings?

Incidental findings are observations on retrieved tissues that may occur during research but that are not related to the goals of research. These observations may reveal susceptibility to developing a particular disease that may be relevant to you and other blood relatives of your child. They will have the chance to decide whether to be informed of prospective incidental findings.

We thank you for the attention and time you have devoted to reading this document. If you decide to allow your child to participate in RTD, please complete and sign the form below. A copy of the information sheet will be given to you.

CONTACTS:
RTD PROGRAM GENERAL INFORMATION   dr. Name, Surname, mobile, email
RTD ETHICIST   dr. Name, Surname, mobile, email

RTD IC2_PG – Parents/Guardians Information Sheet & Informed Consent Form

♣RTD

INFORMED CONSENT FORM

I, the undersigned....................…….. born in………….. on……………

Parent/Guardian of:
name and surname……. born in ……… on……….

Hereby declare to have:

read and understood the information sheet whose contents have been explained by the RTD ethicist
Dr. …………………………;
had time to discuss my decision with other persons;

had time to ask questions and received clarifications on the program.

Therefore, I voluntarily

O   consent   O   refuse

to have my child participate in the RTD program. To this end, I:

- consent to the use of my child's tissues for:

Check the box next to the aim(s) you want your child's tissues to be used for (you can check more than one box)

O   **oncology research only**

O   **research for all diseases**

O   **commercial use of the sample (diagnostic kits, etc.)**

O   **non-medical areas of research (cosmetics, etc.)**

O   **future uses now unknown**

……..…………………..……….Parent 1/legal guardian signature

……..…………………..……….Parent 2/legal guardian signature

…………………………..….…….Signature of the RTD ethicist collecting consent

Date……………….. Place……………

RTD PROGRAM GENERAL INFORMATION   dr. Name, Surname, mobile, email
RTD ETHICIST   dr. Name, Surname, mobile, email

5. RTD – Minors 7-11 Information Sheet & Informed Consent Form

🌿RTD

RTD PROGRAM – INFORMATION SHEET

Dear *name of the boy/girl*

This paper is for you. Read it when you have time and talk about it with your family.
Any questions? We are here for you.

Why is research important?

Research is very important because it provides new knowledge to help people in the future!

RTD is a research program. If you want, you can donate very small parts of your body to scientists!

What is RTD?

Does it hurt?

No!

What good things can happen?

You can help kids like you.

Write your name in capital letters if you want to participate in the study

· · · · · · · · · · · · · · · ·

…………………………………….Signature of the RTD ethicist collecting consent

Date……………….. Place……………

RTD PROGRAM GENERAL INFORMATION   dr. Name, Surname, mobile, email
RTD ETHICIST   dr. Name, Surname, mobile, email

6. RTD – Minors 12-17 Information Sheet & Informed Consent Form

❧RTD

RTD PROGRAM – INFORMATION SHEET

Dear *name of the child,*

As the RTD ethicist, Dr. *insert name ethicist* has previously explained to you, the part of your body that has cancer might help scientists to study your disease and to find treatments in the future to cure children like you. If you want, you can donate your tissues. This program is called Rapid Tissue Donation or RTD, and it is explained in this sheet. Take your time to read this information, if you want, and if you have any questions you ask the RTD ethicist, Dr *insert name ethicist*. Each question is worth asking.

What is Rapid Tissue Donation?
Rapid Tissue Donation (RTD) is a program of tissue donation dedicated to cancer patients like you.
RTD takes place right after death and helps researchers to study the characteristics of cancer tissues to advance research. To this end, it is crucial for tissues to be collected straight after death not to let tissues lose their specific characteristics.

Do I have to participate in RTD?
No, you do not have to participate in RTD. It is your decision. You can discuss this option with your parents, your family, your friends, and your physician. Take your time and think about it. And, if you have any questions at all, please ask the RTD ethicist. The contact details of the RTD ethicist are at the end of this sheet.

What risks are involved for me?
There are no risks for you as tissue collection takes place after death. A physician will take tissues by following rigorous medical and ethical protocols.

What benefits are involved?
There are no direct benefits for you as tissue collection takes place after death. There will be no benefits for your parents and your family. However, your tissues will contribute to the development of cancer research and might benefit future people, including children like you.

What happens if I consent?
If you consent to the program by signing this form, you will be asked again later, just in case you change your mind. You are free to change your mind whenever you want. If you do, call the RTD ethicist, whose contact details are provided at the end of this form. Your care and treatment will not be affected by your choice. You will be supported with the best available treatment and care regardless of your choice.

What happens if I refuse?
If you refuse to enter the program, this is OK: you do not have to sign this form. Nobody will ask again. If you change your mind, and you want information on RTD, you can talk to the RTD ethicist, whose contact details are provided at the end of

this form. If you do not want to enter the program, your care and treatment will not change because of your choice. You will be supported with the best available treatment and care. Your refusal will in no way be interpreted as a lack of trust.

Can I change my mind?

Yes. You can change your mind whenever you want before tissues are collected. It is your right. Just tell the RTD ethicist whose contact details are provided at the end of this sheet.

How will my tissues be used?

Collected tissues will be used for research aims, for example, to understand how cancer appears and grows and to find better treatments to stop it. Research is a complex process that requires time and patience. This means that your tissues can be used for cancer research, for research on other diseases, or to discover things that today we cannot even imagine. For this reason, it is difficult to tell you precisely what your tissues will be used for.

However, we understand that it is important for you to know how your tissues will be used. If you decide to enter RTD, you can choose with your parents the aims for which your tissues will be used. The RTD ethicist will answer all your questions about this.

Who is responsible for the use of my tissues?

The RTD program is responsible for the tissues, which will be managed responsibly and in compliance with ethical and legal frameworks. The RTD program will respect your choices.

Will my tissues be sold for money?

Your tissues cannot be sold for money. However, your tissues can be used for research that will help make commercial products, such as diagnostic kits. But this will not happen if you do not want it to. Just talk to your parents and discuss your preferences. If you allow your tissues to be used for commercial purposes, it will not be possible for you or your family to claim any financial income resulting from these activities.

How long will my tissues be stored?

At the present time, it is not possible to say. Tissues will be kept until their quality is suitable to obtain reliable results.

How are tissues conserved?

In order to safeguard your privacy, tissues will be pseudonymized, which means that they will be associated with a code. This code is unique to you, and only the RTD program knows which code corresponds to which name.

We thank you for your attention and time.
If you decide to participate in RTD, please complete and sign the form below. A copy of the information sheet will be given to you.

CONTACTS:
RTD PROGRAM GENERAL INFORMATION   dr. Name, Surname, mobile, email
RTD ETHICIST   dr. Name, Surname, mobile, email

❧RTD

RTD – INFORMED CONSENT FORM

I, ……………..….. born in………….. on……………

Hereby declare to have:

read and understood the information sheet whose contents have been explained by the RTD ethicist
Dr. ……………………………;
had time to discuss my decision with other persons;

had time to ask questions and received clarification about the program.

Therefore, I voluntarily

O   consent   O   refuse

to participate in the RTD program.

………………………………………Participant signature
………………………………………Signature of the RTD ethicist collecting consent

Date……………….. Place……………

RTD PROGRAM GENERAL INFORMATION   dr. Name, Surname, mobile, email
RTD ETHICIST   dr. Name, Surname, mobile, email

7. RTD IC3–Patient Re -Consent Form

✿RTD

RTD – RE - CONSENT FORM

Dear Patient/Legal Representative/Parent/Guardian,

You are reading this form as you have previously read an information sheet and signed your informed consent form agreeing to participate in Rapid Tissue Donation (RTD).

RTD is a program of tissue donation specifically for cancer patients. It involves the retrieval of cancer tissues and takes place after death. RTD offers researchers the chance to study the characteristics of tumor tissues to understand the development of tumors and to advance research. To this end, it is crucial for tissues to be collected straight after death so that the tissues do not lose their specific characteristics. The window of time required for collection is fixed at two to six hours following death.

The informed consent you signed is attached to this form. Please take your time to read it and to reflect on the choices you made. This form is to confirm, modify, or withdraw your consent.

If you have not changed your mind, you can confirm consent and the choices made on the previous informed consent.

If you have changed your mind, and you no longer want to participate in RTD, you may check the corresponding box.

If you still want to participate but you want to update the choices you made on the previous informed consent form, just check the corresponding box. You will be given a blank copy of the informed consent form in order to update your choices.

If you have any questions, please ask the RTD ethicist whose contact details are provided at the end of this form.

I, the undersigned .... born in............... on...............

Hereby declare to have:

read and understood the information sheet whose contents have been explained by the RTD ethicist Dr. ....;

had time to ask questions and receive clarification about the program;

had time to discuss my decision with other persons;

read the informed consent form previously signed;

reflected on the choices selected within the signed informed consent form.

Therefore, I voluntarily *(check the corresponding box):*

O   confirm consent and the choices made on the previous informed consent;

O   confirm consent but want to update the choices on the previous informed consent;

O   withdraw previous informed consent

....Participant signature

....Signature of the RTD ethicist collecting consent

Next of kin

.... Next of kin name, surname

.... Next of kin signature

Date.................... Place..............

RTD PROGRAM GENERAL INFORMATION   dr. Name, Surname, mobile, email

RTD ETHICIST   dr. Name, Surname, mobile, email

8. RTD – Information Sheet Relatives & Consent Form

♣RTD

INFORMATION SHEET FOR BLOOD RELATIVES OF
A (PROSPECTIVE) RTD DONOR

Dear Sir/Madam,

You are reading this form as one of your blood relatives has signed an informed consent to Rapid Tissue Donation (RTD).

RTD is a program of tissue donation specifically for cancer patients. It involves the retrieval of cancer tissues and takes place after death. RTD offers researchers the chance to study the characteristics of tumor tissues to understand the development of tumors and to advance research. To this end, it is crucial for tissues to be collected straight after death so that the tissues do not lose their specific characteristics. The window of time required for collection is fixed at two to six hours following the death of the patient.

What risks are involved for my relative?

There are no risks for your relative as tissue collection takes place after death. The retrieval of tissues is performed by a physician who specializes in pathology and is assisted by the health care staff. The procedure follows rigorous medical and ethical protocols.

What benefits are involved?

There are no direct benefits for your relative as tissue collection takes place after death. There will be no benefits for your relatives' family. However, tissues will contribute to the development of cancer research and might benefit future patients.

How does RTD work?

The window of time required for tissue collection is fixed at two to six hours following the death of the patient. This means that, when the moment comes, you will be asked after a short period of time to let the retrieval process begin. We know this will be difficult for you and your family. Our staff will be there to support you. The retrieval of tissues is performed by a physician who specializes in pathology and is assisted by the health care staff. The procedure follows rigorous medical and ethical protocols. The RTD ethicist will be in touch with you throughout the retrieval procedure and is responsible for the process.

How long will the process last?

The body will be returned between 24 and 36 hours after the death of the patient. The RTD ethicist will be in touch with you during this process.

How will retrieved tissues be used?

Collected tissues will be used for research aims in order to advance scientific development and translate it into better treatments for patients. Research is a complex and rigorous process that requires time and patience. This means that tissues collected today might reveal themselves useful tomorrow for aims that are not imaginable today. For this reason, it is difficult be precise about the aims for which the tissues will be used.

However, we understand that it is important to know how the tissues will be used. For this reason, we have asked your relative to choose among multiple research areas and to authorize the purposes for which his/her tissues will be used.

Who is responsible for the use of retrieved tissues?

The RTD program is responsible for the retrieved tissues, which will be managed responsibly and in compliance with ethical and legal frameworks. The RTD program will respect your relative's choices.

Will my relative's tissues be commercialized?

Retrieved tissues cannot be sold for profit. However, tissues might be used for research that will lead to commercial products, such as diagnostic kits if your relative has checked the corresponding box. Your relative's family will not be entitled to any financial income resulting from these activities.

How long will retrieved tissues be stored?

At the present time, it is not possible to quantify a precise storage duration, as tissues will be kept until their quality is suitable to obtain reliable results.

How are tissues conserved?

In order to safeguard the privacy of your relative, retrieved tissues will be pseudonymized, which means that they will be associated with a code. Only the RTD program knows which code corresponds to which name.

How will my relative's decision affect me and my other relatives?

If your relative consented to RTD, this might be good to know, as tissues retrieved contain genetic information and research on your relative's tissues might be relevant to his/her blood relatives if incidental findings occur. To this end, you should let the RTD program know if you want incidental findings to be reported to you or not.

What are incidental findings?

Incidental findings are observations on retrieved tissues that may occur during research but that are not related to the goals of research. These observations may reveal a susceptibility to developing a particular disease that may be relevant to blood relatives.

You can decide whether to be reported incidental findings or not. You may also change your mind in the future, by contacting the RTD program in order for them to update your preferences. Contact details for the RTD ethicist can be found at the end of this information sheet.

Your choice will not affect the choice of your other blood relatives.

We know this is a difficult moment for you and for your family.
Our RTD team is here to support you.

8. RTD – Information Sheet Blood Relatives & Consent Form

❧RTD

BLOOD RELATIVE INFORMED CONSENT FORM

I, the undersigned ………………..... born in…….……... on……..……

Contact number……..

Blood Relative of ………

Relation to donor………

Hereby declare to have:

read and understood the information sheet whose contents have been explained by the RTD ethicist Dr. …………………………….;
had time to discuss my decision with other persons;
had time to ask questions and received clarification about incidental findings

Therefore

- In case of incidental findings, I voluntarily declare that:
Check the box next to your choice (you can check only one box)

        O   **I don't want incidental findings to be reported to me**

        O   **I want incidental findings to be reported to** (*please complete*):

………………………………………….Signature
………………………………………….Signature of the RTD ethicist collecting consent

Date……….

RTD PROGRAM GENERAL INFORMATION	dr. Name, Surname, mobile, email
RTD ETHICIST	dr. Name, Surname, mobile, email

Index